Monte Carlo Methods in
Ab Initio Quantum Chemistry

**WORLD SCIENTIFIC LECTURE AND COURSE NOTES
IN CHEMISTRY**

Editor-in-charge: S. H. Lin

World Scientific Lecture and Course Notes in Chemistry — Vol. 1

Monte Carlo Methods in Ab Initio Quantum Chemistry

B. L. Hammond

Computational Research Division, Fujitsu America, Inc.

W. A. Lester, Jr.

Department of Chemistry, University of California, Berkeley

P. J. Reynolds

Physics Division, Office of Naval Research

 **World Scientific**

Singapore • New Jersey • London • Hong Kong

Published by

World Scientific Publishing Co. Pte. Ltd.

P O Box 128, Farrer Road, Singapore 9128

USA office: Suite 1B, 1060 Main Street, River Edge, NJ 07661

UK office: 73 Lynton Mead, Totteridge, London N20 8DH

Library of Congress Cataloging-in-Publication Data

Hammond, B. L.
    Monte Carlo methods in Ab Initio quantum chemistry / B. L. Hammond,
W. A. Lester, Jr., P. J. Reynolds.
      p.   cm. -- (World Scientific lecture and course notes in
chemistry ; vol. 1)
    Includes bibliographical references.
    ISBN 9810203217. -- ISBN 9810203225 (pbk.)
    1. Quantum chemistry.   2. Monte Carlo method.   I. Lester, W. A.
II. Reynolds, Peter J. (Peter James)   III. Title.   IV. Series.
QD462.H29   1994
541.2'8'01519282--dc20   93-47070
                            CIP

Printed in Singapore.

Preface

An accurate and consistent treatment of electron correlation is one of the great challenges currently confronting electronic structure calculations in theoretical chemistry, atomic and molecular physics, and condensed matter theory. Such a treatment is critical for many aspects of the *ab initio* determination of atomic and molecular structure. In these problems, the reference point from which degree of correlation is judged is the Hartree-Fock limit. By convention Hartree-Fock is said to have no correlation since it is a mean field, single particle description. In contrast, the goal of electronic structure theory is the solution of the non-relativistic, Born-Oppenheimer (clamped nuclei) Schrödinger equation. To this end, a number of "post-Hartree-Fock" methods have been developed, i.e. methods that go beyond the mean-field approximation, yet retain the simple molecular-orbital picture. Importantly, these methods add many-body correlation at various levels of approximation. The success of Hartree-Fock and post-Hartree-Fock methods for treating electronic structure benefits considerably from cancellation of errors between two or more computations in the calculation of observables. In many other cases, in particular in the calculation of excited states and classical barrier heights, these errors often do not cancel. Sometimes quantitative accuracy can only be obtained by including vast numbers of terms in various expansions, but such an approach is typically limited in its applicability to small systems.

Post-Hartree-Fock methods fall mainly into two categories: those based on configuration interaction, which expand the wave function in primitive basis functions and determinants (configurations), and those based on many-body perturbation theory which treat electron-correlation as a perturbation. In both cases, correlation corrections usually begin from a Hartree-Fock wave function. Other important formalisms that treat the correlation energy are density functional theory and electron propagator methods. The former approach is rapidly gaining momentum and is useful in a variety of applications.

Our aim in this volume is to present the fundamental background and current status of the application of Monte Carlo to the determination of correlation in electronic structure. We shall use the term *quantum Monte Carlo* (QMC) to refer to methods that directly solve the Schrödinger equation, as contrasted with *variational Monte Carlo* (VMC) methods that evaluate quantum mechanical expectation values by Monte Carlo procedures. QMC differs from post-Hartree-Fock approaches in that it is a fully correlated method from the outset, rather than building on a mean-field approach. We also distinguish QMC here from a large class of other Monte Carlo methods used in the treatment of quantum problems, often also referred to as QMC in the literature. The distinguishing feature of all these QMC methods is their essentially exact nature, and the lack of any special attention that needs to be paid to correlation *per se*. Their differences lie in the class of problems they were designed to treat. Among these other methods we mention Hubbard-Stratanovich approaches designed for lattice models of strongly interacting electrons, and path integral Monte Carlo which is most often encountered in simulations of condensed matter. Both of these latter methods are finite-temperature approaches, important for treating macroscopic systems.

Though the QMC we discuss here also has its roots in condensed matter physics and statistical mechanics, it is a zero-temperature or ground-state approach. As such it is the most appropriate starting point for treating electronic structure by Monte Carlo. Nevertheless, work on adapting other QMC approaches to electronic structure, as well as to adapting the kinds of QMC we discuss here to other realms are currently in progress.

All QMC methods are presently computationally demanding. Nevertheless, as experimental advances push the limits of accuracy required of computational approaches, and as computers continue to achieve greater speed, emphasis must increasingly be placed on the simplicity, consistency and accuracy of a method. QMC holds high promise here. Moreover, Monte Carlo methods are the most natural for massively parallel computation, ultimately providing a great advantage.

The subject matter of this book is divided conceptually into two parts. The first five chapters present the basic concepts in detail. The latter part of the book, chapters 6–9, covers important extensions of these basic methods. Chapter 1 provides background on random numbers, probability distributions, integral estimation, and simulation of simple differential equations. Those readers already familiar with statistics and the use of random numbers may wish to skip this chapter. Chapter 2 treats VMC, describing methods to evaluate the energy and expectation values derived from variational trial wave functions. The power of VMC is that Monte Carlo integration methods enable one to exploit functional forms that are not analytically integrable for many-electron systems. Such forms include those due to Hylleraas as well as more recent forms, all of which depend explicitly on interelectronic distances. In addition, concepts central to both VMC and QMC, such as importance sampling and optimization are introduced here. QMC solution of the Schrödinger equation is the topic of Ch. 3. The focus is on the use of Green's functions to sample the exact ground state energy. Chapter 4 continues the discussion of QMC with consideration of aspects that must be addressed to treat Fermion systems. In Ch. 5 we discuss the important issue of the choice of trial function. Though QMC methods are not based on trial functions or basis set expansions, unlike many other *ab initio* approaches, these functions do play a central role in importance sampling and in the fixed-node method.

These chapters provide the necessary information and methods to evaluate ground state energies of atoms and molecules. The remainder of the book is dedicated to a number of extensions that are necessary for important chemical applications. The determination of excited-states is treated in Ch. 6, while Ch. 7 describes how to evaluate single-state properties, e.g., dipole and quadrupole moments, and multistate properties, such as the transition dipole moment. Chapter 8 discusses the determination of equilibrium geometries and interaction potentials by means of explicit energy differences and through calculation of energy gradients. Finally, in Ch. 9 we explore new directions being developed which allow QMC to more easily treat systems containing heavy atoms.

We have organized the subject matter with the intent that it may be understood at the entering graduate student level while also being of interest to researchers familiar with the topic. A knowledge of quantum mechanics is required, and some knowledge of molecular electronic structure theory is helpful. For senior undergraduate and graduate level courses, the material in the first four chapters is fundamental. If computational projects are to be undertaken, then Ch. 5 should also prove especially useful. The final chapters may be treated as special topics on the basis of time and interest. For those already familiar with electronic structure theory and Monte Carlo methods in general, chapters 2 through 5 will provide background and breadth, whereas the final chapters will be of more topical interest. To facilitate course work and research projects, we have provided explicit algorithms, exercises, suggested reading, and references at the end of each chapter. The algorithms are written using FORTRAN-like syntax; however we have often deviated from FORTRAN in the interests of readability and brevity. In addition, a number of supplementary topics are addressed in the appendices. Conversion factors to and from atomic units are given in Appendix A. Details of the evaluation of determinantal trial functions are discussed in Appendix B. In Appendix C we provide a listing of a FORTRAN77 implementation of diffusion Monte Carlo without importance sampling, which can be used to treat one-dimensional potentials. For those wishing to continue the study of QMC, the electronic structure program QuantumMagiC, developed at the University of California at Berkeley, may be obtained from the Quantum Chemistry Program Exchange (QCPE) at Indiana University, Bloomington, IN 47405 USA. Finally, a comprehensive topical listing of all references used in the text is given in Appendix D.

We wish to thank the many people who have provided input and criticism during the writing of this manuscript. In particular we wish to thank Professor James B. Anderson, Dr. Dario Bressanini, Professor David M. Ceperley, Dr. William Glauser, Professor Harvey Gould, Professor Jamin Adeola Odutola, Dr. Cyrus Umrigar, and WAL's graduate students: Willard Brown, Mark Goodwin, Eric Johnson and Maria Soto. Any mistakes or omissions, however, are entirely our own. We also acknowledge the support of the Department of Energy, the Office of Naval Research, and the SuperComputer Group of Fujitsu America, Inc. for their support during the writing and preparation of the manuscript. This manuscript was typeset with the LaTeX Version 2.09 document preparation system and the Gnuplot Version 2.0 plotting package.

Contents

Chapter 1

Introduction to Monte Carlo Methods

Monte Carlo methods are a class of techniques that can be used to *simulate* the behavior of a physical or mathematical system. They are distinguished from other simulation methods such as molecular dynamics, by being *stochastic*, that is, *non-deterministic* in some manner. This stochastic behavior in Monte Carlo methods generally results from the use of random number sequences. Although it might not be surprising that such an analysis can be used to model random processes, Monte Carlo methods are capable of much more. A classic use is for the evaluation of definite integrals, particularly multidimensional integrals with complicated boundary conditions. The use to which we will apply Monte Carlo is the solution of the well-known partial differential equation, the Schrödinger equation.

Monte Carlo methods are frequently applied in the study of systems with a large number of strongly coupled degrees of freedom. Examples include liquids, disordered materials, and strongly coupled solids. Unlike ideal gases or perfectly ordered crystals, these systems do not simplify readily. The many degrees of freedom present are not separable, making a simulation method, such molecular dynamics or Monte Carlo, a wise choice. Furthermore, use of Monte Carlo is advantageous for evaluating high

1

dimensional integrals, where grid methods become inefficient due to the rapid increase of the number of grid points with dimensionality. Monte Carlo also can be used to simulate many classes of equations that are difficult to solve by standard analytical and numerical methods.

In this chapter we introduce various aspects of statistics and simulation germane to the Monte Carlo solution of the Schrödinger equation. We begin with a discussion of random and pseudorandom numbers in Sec. 1.1. We then present the essentials of Monte Carlo sampling (in Sec. 1.2), integration (in Sec. 1.3), and simulation (in Sec. 1.4). We attempt to discuss only those concepts needed later in the book. Further details may be found in standard statistics texts.[1]

1.1   Random Numbers and Statistical Analysis

A loose definition of a random number is a numerical value resulting from a process or experiment whose value cannot be predetermined by the initial conditions. It is important to note that the term "random number" is somewhat misleading; a number is not random, rather it is the relationship between numbers in a set which is random. Many natural processes display randomness -- from the decay of subatomic particles to the trajectories of dust particles across the surface of a liquid.

We need to start by defining the concepts and notation used to discuss random numbers and events. An *experiment* is the process of observing one or a set of physical properties in a system of interest. The result of an experiment is limited to certain values or ranges of values of the physical properties. A *state* is an allowed value of the set of physical properties of the system. The set of all possible states is the *sample space*. A *discrete* sample space contains either a finite or infinite number of distinct values. A *continuous* sample space contains an infinite number of continuous values (such as the positions of particles). A *sample point* is a single point in sample space.

A *random variable*, or *variate*, is a variable whose value lies within the sample space with a certain probability distribution. To avoid confusion, we will use upper case (X, Y, Z) to denote sample points and lower case (x, y, z) to denote variables. This distinction will become clear with usage. A *sequence* is a series, in order of occurrence, of sample points resulting from an experiment. We often will use the set notation $\{X_i\}$ to denote all the members of a sequence.

1.1.1 Probability density and distribution functions

The most familiar uses of random numbers occur in games of chance. This connection gives the Monte Carlo method its name. Consider a standard six-sided die. If one tosses an ideal, unbiased die, and records the outcome for a sufficiently large number of tosses (in principle, an infinite number), each of the six outcomes will occur exactly one sixth of the time. Even though the outcome of a single toss is random, and thus unknown beforehand, the probability of each outcome is 1/6. The *probability density function* is the function that describes the probabilities of all possible events. The sum or integral of the probabilities must be unity to insure the proper normalization of the density function. For a discrete distribution the normalized probability function p must satisfy,

$$\sum_{i=1}^{N} p(x_i) = 1, \tag{1.1}$$

where the sum is over all states, x_i. In the case of the die, the normalized probability density function is $p(x_i) = 1/6$, for each $x_i = 1, 2, 3, 4, 5, 6$. For the general one-dimensional case, the discrete density function can be represented by a histogram as in Fig. 1.1. This figure illustrates the probabilities of the various sums of two dice. A simple discrete probability density function, such as the one shown in Fig. 1.1, can be determined combinatorially by counting the occurrence of each possible event. In many physical situations, however, the probability density function must be deter-

Figure 1.1: Histogram of the probability density function of the sum of two dice.

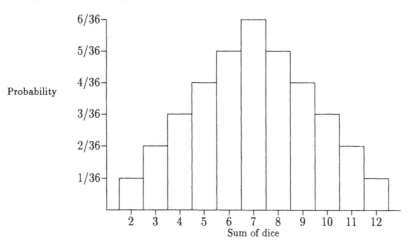

mined by *binning*, because there are too many states to enumerate.

In binning one divides the states into bins representing ranges of states, then performs an experiment tallying the number of instances of each state in the corresponding bin. The count in each bin is normalized by dividing by the total number of counts. If the total number of counts is sufficiently large, the resulting histogram will represent the true density function, such as in Fig. 1.1.

A continuous density function describes the probability of an event associated with a continuous variable. In this case $p(x)dx$ is the probability of an event occurring between x and $x + dx$. For continuous probability density functions, the probability that an outcome will be exactly x is zero.

The *probability distribution function* represents the probability that the value of a given sample point is less than or equal to x, i.e.

$$P(x) \equiv \int_{-\infty}^{x} p(y)dy. \tag{1.2}$$

(We will use lower case to denote density functions and upper case to denote the associated distribution functions.) The distribution function always increases monotonically from zero to one. Let us illustrate these concepts by examining two cases that will be of importance in later chapters: the *uniform* and *Gaussian* distributions (see Figs. 1.2 and 1.3). The density function of the uniform distribution is illustrated in Fig. 1.2. For the uniform distribution, denoted by $u(x)$, all outcomes in a given range $[\alpha, \beta]$ have equal probability, and all other values of x have zero probability. The normalization of u (Eq. 1.1 with integration replacing the sum) requires that

$$u(x) = \begin{cases} (\beta - \alpha)^{-1} & \alpha \leq x \leq \beta \\ 0 & \text{otherwise.} \end{cases} \tag{1.3}$$

Therefore the probability that x is between a and b if $\alpha \leq a < b \leq \beta$ is given by

$$\int_{a}^{b} u(x)dx = U(b) - U(a) = (b - a)/(\beta - \alpha), \tag{1.4}$$

where U is the distribution function associated with u.

The Gaussian probability distribution function, $g(x)$, shown in Fig. 1.3, owes much of its importance to the central limit theorem (discussed in Sec. 1.1.3). In one dimension its density function is

$$g(x) = \frac{e^{-(x-\mu)^2/2\sigma^2}}{\sqrt{(2\pi\sigma^2)}}, \tag{1.5}$$

where the parameter μ specifies the center of the density function and σ determines its width. The probability of observing a number x in the interval a to b is given by the integral

$$\int_{a}^{b} g(x)dx = G(b) - G(a). \tag{1.6}$$

Figure 1.2: Probability density function of the uniform distribution.

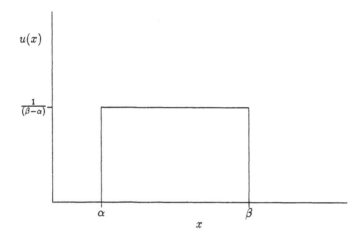

The integral of Eq. 1.6 is not known in closed analytical form, but it can be related by a change of limits to the error function (Exercise 1),

$$\operatorname{erf}(y) \equiv \frac{2}{\sqrt{\pi}} \int\limits_{0}^{y} e^{-x^2} dx, \tag{1.7}$$

which is found in standard tables.[2]

The above discussion has been for probability density functions of a single variable. A *multivariate* probability density function $p(x_1, x_2, \cdots, x_n)$ gives the probability of the state $x_1 = X_1, \cdots, x_n = X_n$ (the discrete case), or $x_1 \leq X_1 \leq x_1 + dx_1, \cdots, x_n \leq X_n \leq x_n + dx_n$ (the continuous case). The arguments $(x_1, \cdots, x_n)$ are conveniently denoted by the n-dimensional vector $\mathbf{x}$. Because we will be concerned only with continuous distributions in later chapters, we will restrict further discussion to this case. To obtain appropriate equations for discrete distributions one simply replaces integrals by sums over states.

Figure 1.3: Probability density function of the Gaussian distribution.

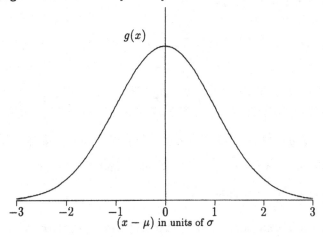

Multivariate probability density functions are normalized in the usual way, namely,

$$\int\limits_{-\infty}^{\infty} p(\mathbf{x})d\mathbf{x} = 1. \tag{1.8}$$

Just as in the single variable case, we can find the probability that component x_1 is in the interval a_1 to b_1 while x_2 is between a_2 and b_2, etc.,

$$P(a_1 \le x_1 \le b_1, a_2 \le x_2 \le b_2, \cdots, a_n \le x_n \le b_n) = \int\limits_{a_1}^{b_1}\int\limits_{a_2}^{b_2} \cdots \int\limits_{a_n}^{b_n} p(\mathbf{x})d\mathbf{x}. \tag{1.9}$$

In addition, for multivariate probability density functions, we can define a *marginal density*, p_i, which is the probability that a single component lies within the range $x_i \le X_i \le x_i + dx_i$, regardless of the values of the other components, i.e.

$$p_i(x_i) = \int\limits_{-\infty}^{\infty} p(\mathbf{x})dx_1 dx_2 \cdots dx_{i-1} dx_{i+1} \cdots dx_n. \tag{1.10}$$

Marginal densities are of interest whenever one is interested in the properties of a single variable or set of variables. The quantum mechanical concept of a one-electron density in a many-electron system is an example of a marginal density function.

The bivariate Gaussian distribution serves as a good example of a multivariate probability density function. Its form is given by

$$g(x_1, x_2) = \frac{e^{-Q/2}}{2\pi\sigma_1\sigma_2\sqrt{1 - \rho^2}} \qquad (1.11)$$

where

$$Q = \frac{1}{1 - \rho^2}\left[\frac{(x_1 - \mu_1)^2}{\sigma_1^2} - 2\rho\frac{(x_1 - \mu_1)(x_2 - \mu_2)}{\sigma_1\sigma_2} + \frac{(x_2 - \mu_2)^2}{\sigma_2^2}\right]. \qquad (1.12)$$

The quantities σ_1 and σ_2 determine the widths of the distribution along the x_1 and x_2 axes, and ρ depends on the correlation between x_1 and x_2, which is a skewing of the distribution away from the x_1 and x_2 axes.

1.1.2   Characterization of probability density functions

A probability density function may be characterized by a number of different measures. The *expected value* of an arbitrary function, $f(x)$, with respect to a probability density function, $p(x)$, is given by

$$\langle f \rangle_p \equiv \int\limits_{-\infty}^{\infty} f(x)p(x)dx. \qquad (1.13)$$

The bracket notation $\langle \cdots \rangle_p$ denotes the average value with respect to the density p. When the subscript p is omitted, then the density function is assumed to be readily identifiable from the context. This notation is distinct from the Dirac bracket notation used in later chapters.

An often-important function is $f(x) = x^k$ whose expectation value defines the k-th *moment* of the distribution,

$$\langle x^k \rangle_p \equiv \int\limits_{-\infty}^{\infty} x^k p(x)dx. \qquad (1.14)$$

The first moment, $\langle x \rangle$, is the *mean* value of x over $p(x)$, sometimes denoted as $\bar{x}$. A probability density function is uniquely determined by all of its moments (in principle, an infinite number). The k-th *central* moment is the k-th moment about the mean,

$$\langle (x - \bar{x})^k \rangle_p \equiv \int_{-\infty}^{\infty} (x - \bar{x})^k p(x) dx. \tag{1.15}$$

The most frequently used of the central moments is the second moment or *variance*, denoted by σ^2. The square-root of the variance is the *standard deviation*, σ, The variance can be evaluated from Eq. 1.15, which would require the pre-calculation of $\bar{x}$. This pre-calculation can be avoided by manipulating Eq. 1.15 (for $k = 2$) to obtain,

$$\sigma^2 = \langle x^2 \rangle - \bar{x}^2, \tag{1.16}$$

which enables one to accumulate $\langle x^2 \rangle$ and $\bar{x}$ simultaneously during an experiment. Using our previous examples, the mean of the uniform distribution u is

$$\langle x \rangle_u = \int_{\alpha}^{\beta} x u(x) dx = \frac{1}{2}(\beta + \alpha), \tag{1.17}$$

and the corresponding variance is

$$\sigma_u^2 = \int_{\alpha}^{\beta} x^2 u(x) dx - \langle x \rangle_u^2 = \frac{1}{12}(\beta - \alpha)^2. \tag{1.18}$$

(Exercise 1 examines the moments of the Gaussian distribution.)

For the multivariate case, the mean, variance, and higher moments of a particular component, x_i, are defined as above, and denoted as $\bar{x}_i$, σ_i^2, etc. In addition, there also are multivariate moments involving two or more components. The most used of these is the *covariance* (which is a bivariate central moment) of variables x_i and x_j, namely,

$$\mathrm{Cov}(x_i, x_j) \equiv \langle (x_i - \bar{x}_i)(x_j - \bar{x}_j) \rangle. \tag{1.19}$$

The variances of the individual x_i are thus the diagonal elements of the covariance matrix. The off-diagonal elements are a measure of the correlation between two variables. (See Exercise 2 for a simple example using the bivariate Gaussian distribution.)

Often we wish to know whether a set of variables is *independent* or *dependent*. That is, can each variable be chosen from a separate distribution or is the probability of choosing one affected by the outcome of another? This concept will be of central importance when we seek to generate sequences of independent random coordinates for use in simulations. The variates $x_1, \cdots, x_n$ are *independent* if their joint probability density function is the product of univariate probability density functions, i.e. if

$$p(\mathbf{x}) = p_1(x_1)p_2(x_2)\cdots p_n(x_n), \tag{1.20}$$

where $p_i(x_i)$ is the probability density function for component x_i. If two random variables, x_i and x_j are independent, then no correlation exists between them, and therefore $\langle x_i x_j \rangle = \langle x_i \rangle \langle x_j \rangle$ and $\text{Cov}(x_i, x_j) = 0$ (see Fig. 1.4a). If two variables are dependent (as in Fig. 1.4b), one variable can be formulated as a function of the other, e.g., $x_j = f(x_i)$, and their covariance can be either positive or negative.

From Eq. 1.20 it can be shown that independent variables and their probability density functions obey the following relationships:

1. The p_i are normalized.

2. p_i is the marginal density of x_i (i.e. Eq. 1.10).

3. $\langle x_i \rangle_p$ is equal to $\langle x_i \rangle_{p_i}$.

4. σ_i of $p(\mathbf{x})$ is equal to σ of p_i.

5. The covariance between any two variables $(i \neq j)$ is zero.

There is an important caveat here. The converses of these statements are not necessarily true. For example, although $\text{Cov}(x_i, x_j) = 0$ when x_i and x_j are independent, a zero covariance does not imply independence of two variables.

Figure 1.4: Comparison of (a) independent and (b) dependent variables.

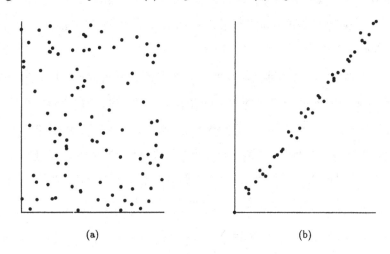

(a)   (b)

Equation 1.13 shows how to compute averages with respect to a given density function. Let us now consider the more common experimental case where the probability density function is not known, but rather a sequence of M numbers, $\{X_i\}$, is given as the outcome of an experiment. One could measure the density function by binning the results as discussed in Sec. 1.1.1. A more powerful approach is available, however. If the $\{X_i\}$ are distributed according to the function $p(x)$, one can readily obtain the average of an arbitrary function f over p as

$$\langle f \rangle_p = \lim_{M \to \infty} \frac{1}{M} \sum_{i=1}^{M} f(X_i). \tag{1.21}$$

(We will explore the consequences of the limit $M \to \infty$ in Sec. 1.1.3.) From our earlier discussion we see that e.g. we can obtain the various moments of p. For two probability density functions to be equal, each of their moments must be equal. Without the analytic form of each probability density function however, this comparison is difficult to make, especially since the variance of the higher moments tends to increase with the power k of the moment. Rather one can ask: what is the probability that the measured probability density function is equal to a known probability density function?

One way to answer this question is by the χ^2 (Chi-squared) test.[3] Suppose we have done the above-mentioned binning procedure, and have obtained n_i, $i = 1, N$, instances in each bin, and we wish to compare this to the known density function p. Let the total number of points be n_{tot}. Then we can judge the quality of the fit between the experimental data and the expected number in bin i, $n_{tot}p(x_i)$, by measuring the weighed squares of the differences, i.e.

$$\chi^2 \equiv \sum_{i=1}^{N} \frac{(n_i - n_{tot}p(x_i))^2}{n_{tot}p(x_i)}. \tag{1.22}$$

If the underlying density function and p match, then n_i and $n_{tot}p(x_i)$ should be equal for $n_{tot} \to \infty$, and the differences, $n_i - n_{tot}p(x_i)$, represent measurement errors. If we discount systematic errors, such measurement errors are themselves random numbers with their own probability density function. Most often such errors are distributed according to a Gaussian function (a consequence of the central limit theorem, Sec. 1.1.3). If this is the case, then the quantity χ^2 is distributed by the *Chi-squared* distribution. This probability distribution function is

$$P(\chi^2|\nu) = \left[2^{\frac{\nu}{2}}\Gamma\left(\frac{\nu}{2}\right)\right]^{-1} \int_0^{\chi^2} e^{-\frac{t}{2}} t^{(\frac{\nu}{2}-1)} dt, \tag{1.23}$$

where Γ is the gamma function, χ^2 is a random variable, and ν is the "number of degrees of freedom." The proper value of ν, assuming no constraints have been placed on the data, is $N-1$. Once the value of χ^2 has been computed by Eq. 1.22, the probability that the two distributions are the same can be obtained from standard statistical Chi-squared tables.[2] In Table 1.1 values of $Q(\chi^2|\nu) \equiv 1 - P(\chi^2|\nu)$ for

Table 1.1: Values of the χ^2 complementary distribution function, $Q(\chi^2|\nu)$.

χ^2	$Q(\chi^2|10)$	$Q(\chi^2|20)$	$Q(\chi^2|30)$
3.0	0.98142	0.99999	0.99999
3.8	0.95592	0.99997	0.99999
6.0	0.81526	0.99890	0.99999
8.0	0.62884	0.99187	0.99998
10.0	0.44049	0.96817	0.99977
12.0	0.28506	0.91608	0.99860
15.0	0.13206	0.77641	0.98974
20.0	0.02925	0.45793	0.91654
25.0	0.00535	0.20143	0.72503
30.0	0.00086	0.06985	0.46565
35.0	0.00012	0.02010	0.24264
40.0	0.00002	0.00500	0.10486

a given ν and χ^2 are listed. The value of $Q(\chi^2|\nu)$ has the following meaning: for large n_{tot}, $Q(\chi^2|\nu)$ is the probability that the data should have a χ^2 greater than the measured χ^2.

As an example, suppose we toss a pair of dice 108 times and record the following data (in x_i, n_i pairs):

(2,1), (3,6), (4,8), (5,10), (6,13), (7,19), (8,18), (9,15), (10,10), (11,5), (12,3).

The histogram of the toss of an ideal pair of dice was given in Fig. 1.1. For the above data $\chi^2 = 3.73$. There are 11 bins and from Table 1.1 $Q(\chi^2|10)$ is approximately 0.96. This result means there is a 4% chance χ^2 would be smaller than the computed value and a 96% chance that χ^2 should be greater than this value. Hence, if this χ^2 is confirmed by further data, we may conclude that the above data is not the result of the toss of fair die! For large n_{tot} we would expect our measured value to have a $Q(\chi^2|\nu)$ in a central range, say between 0.25 and 0.75, corresponding in this example to a χ^2 value between 6.7 and 12.5. In Sec. 1.2.1 we will use the χ^2 test as one measure of randomness when we address the issue of generating random numbers from a desired distribution.

1.1.3 Functions of random variates and the central limit theorem

A function of random variables is itself a random variate. Its mean, variance and other expected values can all be evaluated from its moments $\langle f^k \rangle_p$ and $\langle (f - \bar{f})^k \rangle_p$. An important special case arises when f is a linear function,

$$f(\mathbf{x}) = \sum_{i=1}^{n} c_i x_i. \tag{1.24}$$

In this case the mean is given by

$$\langle f \rangle_p = \sum_{i=1}^{n} c_i \langle x_i \rangle_p, \tag{1.25}$$

and the variance can be shown to be

$$\sigma_f^2 = \sum_{i=1}^{n} c_i{}^2 \sigma_i{}^2 + 2 \sum_{i>j=1}^{n} c_i c_j \mathrm{Cov}(x_i, x_j). \tag{1.26}$$

We are now in a position to obtain the expectation values and variances from a set of experimental data using Eq. 1.21 and Eqs. 1.25–1.26. In Eq. 1.21 we took the limit $M \to \infty$; here we consider the realistic situation where there is only finite information. Suppose we perform an experiment in which we measure a sequence of M independent random outcomes $\{X_i\}$ that are distributed according to the (unknown) density function $p(x)$. The standard *estimate* of the mean value of x is the sum,

$$S(X_1, \cdots, X_M) \equiv \frac{1}{M} \sum_{i=1}^{M} X_i. \tag{1.27}$$

The true value, $\bar{x}$, is obtained only in the limit that $M \to \infty$. Suppose further that we continue the experiment until we have obtained N independent sums, S_1 to S_N *each* composed of M samples. The *central limit theorem*[1] yields the distribution of the

S_i. Such sums, composed as in Eq. 1.27 of random numbers distributed according to *any* probability density function (with bounded second moment), will be distributed according to a Gaussian probability density function whenever M is large. Thus we can write

$$\bar{x} = \frac{1}{N} \sum_{i=1}^{N} S_i. \qquad (1.28)$$

Note that Eq. 1.28 is of the form of Eq. 1.24. Thus we may use Eq. 1.26 to estimate the *variance of the mean* of a finite sample, in this case S, denoted by $\sigma_{\bar{x}}^2$, which is the variance of the S_i, not of the underlying probability density function from which the $\{X_i\}$ were drawn. The variance of the mean provides a measure of the certainty with which one of the estimates S_i is equal to $\bar{x}$. It is safe to assume that the partial sums S_i are independent samples of $\bar{x}$ (which is true for large M) and therefore that the covariance term in Eq. 1.26 vanishes. By the central limit theorem these sums are Gaussian distributed with variance denoted by σ^2. Comparing Eq. 1.27 and Eq. 1.24 we see that $n = N$ and $c_i = 1/N$. Thus from Eq. 1.26 we find that the variance of our estimate of $\bar{x}$ is

$$\sigma_{\bar{x}}^2 = \frac{\sigma^2}{N}. \qquad (1.29)$$

Equation 1.27 is strictly true only if $\bar{x}$ is known exactly. Because $\bar{x}$ is in reality measured, as well as $\sigma_{\bar{x}}^2$, the denominator N is properly reduced to $N - 1$, yielding the familiar expression given in statistics texts,

$$\sigma_{\bar{x}}^2 = \frac{\sigma^2}{N - 1}. \qquad (1.30)$$

For the Gaussian distribution, the probability of a measured result falling between $\pm \sigma_{\bar{x}}$ of the exact $\bar{x}$ is ~ 0.68; the probability for falling within $\pm 2\sigma_{\bar{x}}$ is ~ 0.95. Therefore, if M is sufficiently large for the central limit theorem to hold, then the sums S_i in the above experiment provide an estimate of $\bar{x}$ that has a 68% chance of being within $\sigma_{\bar{x}}$, and a 95% chance of being within $2\sigma_{\bar{x}}$, of the true mean.

1.2   Generation of Pseudorandom Numbers

Let us now turn to the generation of random sequences of numbers drawn from a given probability density function. There is no algorithm (i.e., procedure based on mathematical operations) for generating true random sequences, since an algorithm implies a deterministic relationship between the numbers. Truly random sequences result from certain kinds of physical processes, e.g. radioactive decay. In the past, large lists of random numbers derived from physical data were tabulated for use in Monte Carlo. Such lists, however, were cumbersome and inefficient to use. Alternatively, we could build a device based on a natural process to supply numbers as needed. Generally the rate of generation is a limitation, though this approach has been used on occasion. Moreover, one would prefer a method that can reproduce the same sequence of random numbers when it is necessary to check the consistency and accuracy of the produced results.

One approach is to use a *pseudorandom* number generator, obtained from a prescribed set of mathematical operations, or algorithm, which thus yields a reproducible sequence of numbers. The resulting pseudorandom numbers are not truly random, but they do possess certain attributes of true random numbers sequences. The most crucial test of a pseudorandom number generator is that the mean and variance of the number sequence produced match those of the desired probability density function. Ideally, higher moments also should match. A number of other tests have been designed to measure the quality of a pseudorandom number generator. These tests are discussed together with pseudorandom number generation in the following section.

1.2.1   Generation of uniform variates

Besides the many physical examples in which uniform variates come into play, they also are important because the generation of *arbitrary* density functions usually begins by transforming from a uniform distribution. Thus it is important to first focus on these uniform variates. Our approach will be via pseudorandom number generators.

We will take some time to discuss linear congruential generators, which are the most commonly used class of uniform pseudorandom number generators. The basic idea is to generate integers between 0 and $m - 1$, where m is very large, in a random sequence. Typically m is 2^b where b is the number of bits in an integer on the computer used. One then divides these integers by $m - 1$ to obtain real numbers between zero and one. To start the procedure off, one chooses an integer between 0 and $m - 1$ to be the generator *seed*, I_0. The seed is the starting value of the pseudorandom number sequence. The same seed always produces the same sequence. Once the seed is specified, the next integer, or in general, the $j + 1$ integer in the sequence, denoted by I_{j+1}, is obtained from I_j by

$$I_{j+1} = (aI_j + b)\text{modulo } m. \qquad (1.31)$$

Equation 1.31 defines the linear congruential generator. Division by $m - 1$ scales these integers onto the desired interval. As we will see, the quality of the generator depends upon the choice of a, b, and m.

Much has been written on the advantages and disadvantages of the linear congruential generator. The main advantage is that it can produce numbers quite uniformly between 0 and 1 with a minimum number of mathematical steps. There also are some well-documented faults. First, clearly not all numbers between 0 and 1 are possible outcomes, only multiples of $(m - 1)^{-1}$. This is the *grain size*. On digital computers this error can be made less than the truncation error due to a finite word length by choosing m to be 2^b as described above. A more serious problem is that Eq. 1.31

is cyclic and will return to the seed I_0 after P steps, its period length. Hence the entire sequence will repeat at that point. The maximum period length is m. For good choices of a and b, the series will be of maximal length. Even this maximum series length can limit the precision of Monte Carlo results if more random points are needed than the series length allows.

Maximum series length does not guarantee randomness. One could, e.g., choose the trivial case, $a = 1$ and $b = 1$. The sequence is m long, but is clearly not random. Another difficulty occurs when the generated pseudorandom numbers are not used individually, but as k-tuples, as in the Cartesian coordinate vector $(x_1, x_2, \cdots, x_k)$. It has been shown that numbers generated by Eq. 1.31 will not fill k-dimensional space uniformly, but rather they will lie on $k - 1$-dimensional hyperplanes. Again, depending on a, b, and m, there can be a maximum of $m^{1/k}$ such hyperplanes, and significantly fewer hyperplanes for poor choices of parameters.

Based on a theoretical analysis of the linear congruential generator[3] the criteria for the best choice of a and b are:

1. b is a prime number close to $m(\frac{1}{2} - \frac{\sqrt{3}}{6})$;

2. $a - 1$ is a multiple of every prime factor of m;

3. $a - 1$ is a multiple of 4 if m is a multiple of 4.

The proof of these criteria, and the mathematics associated with predicting the properties of linear congruential generators, is very involved. For the purposes of using such generators we will focus on tests to detect problems with the generated series. Such tests are useful regardless of how the random sequences are generated. We will concentrate on the χ^2 test, the serial correlation test, and the k-tuples test. We will apply these tests on the sequence $\{X_i\}$, $0 \le i \le N$.

THE χ^2 TEST: We have already described the χ^2 test in Sec. 1.1.2. To test the distribution produced by a given value of a, b and m, one can bin a large sequence of numbers and apply the χ^2 test as discussed in Sec. 1.1.2. This test should be done several times with independent sequences to insure that the results are valid.

SERIAL CORRELATION TEST: We desire the points in the sequences produced to be independent. However, Eq. 1.31 shows that each point depends explicitly upon the previous point. To determine the extent of serial correlation, let us define two overlapping subsequences $\{U\}$ and $\{V\}$. $\{U\}$ consists of all the points 0 to $N - 1$, and $\{V\}$ consists of all the points 1 to N. The serial correlation is measured by the correlation coefficient

$$r = \frac{\text{Cov}(U, V)}{\sigma^2(X)}. \tag{1.32}$$

The quantity r is a measure of the correlation between U_i and U_{i+1}. It ranges from -1 to $+1$, where a value near zero indicates that serial correlation is not present, and a value of ± 1 indicates total linear dependence. Even with independent points, for any finite N, it is important to note that r will not be exactly zero. One could also measure the correlation between any other shifted subsequences, e.g. by defining $\{U\}$ to be the points 0 to $N - j$ and $\{V\}$ to be j to N in order to measure the correlation between U_i and U_{i+j}.

THE k-TUPLES TEST: In the remainder of this book we shall require random vectors related to the positions of electrons. We should therefore apply the above tests to the subsequences formed by *each* component of the various vectors. For example, if we create N random vectors of dimension k, i.e. $\{\mathbf{X}_i\} \equiv \{(X_{i_1}, X_{i_2}, \cdots, X_{i_k})\}$, then the subsequences $\{X_{i_\lambda}\}$, should pass all tests for randomness for each component λ.

1.2.2   Generation of non-uniform random variates

Frequently, one requires a distribution other than the uniform one. There are many probability density function-specific tricks one can devise to generate particular distributions. We describe instead two general approaches and their limitations: the *inversion* method and the *rejection* method.

In the inversion method one transforms a uniform distribution of pseudorandom numbers into the desired probability density function as follows. Suppose x is a random variate selected from the uniform density function $u(x)$. Let y denote a random variate selected from the desired density function $w(y)$. We wish to find a function, $f(x)$, that transforms x into y, i.e. $f(x) = y$. This is a well-known problem in probability theory. It can be shown that,[1]

$$w(y) = w(f(x)) = u(x)\left|\frac{dx}{dy}\right| = u(f^{-1}(y))\left|\frac{df^{-1}(y)}{dy}\right|. \tag{1.33}$$

If we rearrange Eq. 1.33 and integrate over y (using the fact that u is normalized), we obtain,

$$f^{-1}(y) = \int_{-\infty}^{y} w(y')dy' = W(y). \tag{1.34}$$

Hence we see that $f(x) = W^{-1}(x)$, where W is the probability distribution function of $w(y)$. The random numbers, y, are therefore generated by finding the function $f(x)$ that is the inverse function of W. Note that the use of this procedure requires that the integral of $w(y)$ is known and invertible.

Consider for example the exponential distribution, with density $w(y) = ke^{-ky}$ in the range $0 \leq y \leq \infty$, and zero elsewhere. This distribution governs the temporal behavior of first-order rate processes with rate constant k. This distribution will also be used later in the solution of the Schrödinger equation. The function $W = \int_0^y ke^{-ky'}dy' = 1 - e^{-ky}$ can be readily inverted to yield $f(x) = -k\ln(1-x)$. Because

$1 - x$ is uniformly distributed if x is, we may omit the 1 to yield,

$$y = -k \ln x. \tag{1.35}$$

Exponentially distributed numbers in the range $0 \leq y \leq \infty$ are then generated by first generating a uniformly distributed random variate, X, between zero and one, and then calculating $Y = -k \ln X$.

Another useful application of the inversion method is the Box-Muller[4] algorithm for generating a Gaussian distribution. Straightforward application of the inversion method to the one-dimensional Gaussian distribution fails because the required indefinite integral is not known in closed form (although a lookup table could be used). However, this integral *is* known for the two-dimensional case. Thus it turns out that we can generate *pairs* of Gaussian random numbers.

For a two-dimensional distribution, suppose we have a function $w(y_1, y_2)$ which is the joint probability density function for y_1 and y_2 (see Sec. 1.1.1). The uniformly distributed variates (x_1, x_2) can then be transformed to the desired set (y_1, y_2) by a relation analogous to Eq. 1.33,

$$w(y_1, y_2) = u(x_1, x_2)\mathbf{J} \tag{1.36}$$

where $\mathbf{J}$ is the Jacobian,

$$\mathbf{J} \equiv \det \begin{bmatrix} \frac{\partial x_1}{\partial y_1} & \frac{\partial x_2}{\partial y_1} \\ \frac{\partial x_1}{\partial y_2} & \frac{\partial x_2}{\partial y_2} \end{bmatrix}. \tag{1.37}$$

To generate Gaussian random variates, the Box-Muller method uses the two-dimensional probability density function $w(y_1, y_2) = \exp(-y_1^2/2)\exp(-y_2^2/2)$. The two variables y_1 and y_2 are independent Gaussian random variates since for this form of w, $\mathrm{Cov}(y_1, y_2)$ vanishes (see Exercise 2). Transforming to polar coordinates, r and θ, using $y_1 = r\cos\theta$ and $y_2 = r\sin\theta$, the distribution function becomes

$$W(y_1, y_2) = \int_{-\infty}^{y_1} \int_{-\infty}^{y_2} w(y_1', y_2') dy_1' dy_2'$$

$$= \int_0^r \int_0^\theta r' e^{-r'^2/2} dr' d\theta'$$
$$= \theta(1 - e^{-t}), \tag{1.38}$$

where $t = r^2/2$. The desired transform is the inverse of Eq. 1.38, which becomes (upon considering $\mathbf{J}$) $2\pi x_1 x_2 = \theta e^{-t}$. Thus we can choose x_1 as,

$$x_1 = e^{-t} = e^{-(y_1^2 + y_2^2)/2} \tag{1.39}$$

and x_2 as

$$x_2 = \theta/2\pi = (2\pi)^{-1} \arctan(y_2/y_1). \tag{1.40}$$

Pseudorandom Gaussian numbers are then generated in pairs by the inverse of Eq. 1.39 and Eq. 1.40, namely,

$$y_1 = \sqrt{-2\ln x_1} \cos(2\pi x_2) \tag{1.41}$$

$$y_2 = \sqrt{-2\ln x_1} \sin(2\pi x_2). \tag{1.42}$$

If only a single pseudorandom Gaussian is required at a time, then the other may be stored for later use. So far we assumed that the mean and variance of the Gaussian are zero and one, respectively (see below Eq. 1.37). To generate random numbers with mean μ and variance σ^2, one can simply transform to new variables $\tilde{y}_i$ related to y_i as $y_i = (\tilde{y}_i - \mu)/\sigma^2$, $i = 1, 2$. This results in

$$\tilde{y}_i = \sigma^2 y_i + \mu. \tag{1.43}$$

Because the inversion procedure has the drawback that the distribution function must be known and invertible, we next present a more widely applicable method to generate a desired probability density function — the rejection method. To implement the rejection method one first finds a function, $h(y)$, that is everywhere greater than (but preferably close to) the desired probability density function, $w(y)$, and for which a pseudorandom number generator exists. The simplest choice is a uniform distribution over the domain of w with $h(y) = w_{max}$, the maximum value of $w(y)$. Then random number sequences are generated using the uniform distribution, $u(y)$ scaled by w_{max}. This situation is shown in Fig. 1.5. For a given value of y, one needs to adjust $h(y)$ to $w(y)$. This can be done by first generating a uniform random variate Y, then forming the ratio $p(Y) \equiv w(Y)/h(Y) \leq 1$. The number Y is accepted as a member of the w distribution with probability p. This last step involves generating a second uniform random number, ξ, between 0 and 1, and *rejecting* Y if $\xi > p$. The effect of the rejection is to weight h by the ratio $p = w/h$, so that w results. In general, one wishes h to mimic w as closely as possible to minimize the rejection of computed pseudorandom numbers. This method is simple and powerful, but lacks efficiency if h is not close to w. Loss of efficiency will be particularly important for high-dimensional spaces. In the next sections we will develop a more efficient form of the rejection method known as Metropolis sampling.

1.3 Random Walks and Metropolis Sampling

Armed with pseudorandom numbers and statistical analysis methods, we are now ready to generate general distributions — necessary for the treatment of many physical problems — by applying the Monte Carlo method. One of the most powerful tools within Monte Carlo is the *random walk*. While simple random walks can be used directly, to model physical processes such as diffusion and particle showers, a generalized walk can be used to generate most desired distributions.

Perhaps the most famous random walk is the "drunkard's walk," in which a walker takes random moves (from corner bar to corner bar) on a two-dimensional lattice. This walk is readily generalized to higher dimensions and to continuous space. In this section we shall focus on developing a random walk method that enables one to

Figure 1.5: Illustration of the rejection method.

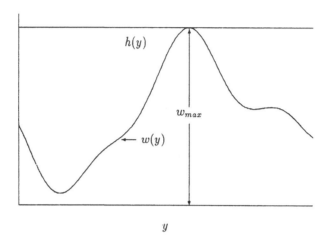

sample pseudorandom numbers from general multivariate probability density functions. The method we discuss was originally introduced by Metropolis, Rosenbluth, Rosenbluth, Teller and Teller[6], referred to herein as Metropolis sampling. It is also sometimes labeled the $M(RT)^2$ algorithm (after the unique initials of the authors). This algorithm has been widely applied to the generation of multivariate probability density functions, particularly in statistical mechanics and more recently in quantum Monte Carlo algorithms. We begin with some necessary preliminaries.

1.3.1   Markov chains

In a random walk one defines a mathematical entity, called a walker, whose attributes completely define the state of the system in question. The state of a system can refer to any physical quantities, from the vibrational states of a molecule specified by a set of quantum numbers, to the brands of coffee available to a consumer in a supermarket.

The walker moves in an appropriate state space by a combination of deterministic and random displacements from its previous position. This sequence of steps forms a *chain*. Consider a discrete system for which there are N available states, denoted S_1 through S_N. Let the state of the system at a discrete point in time i be $x^{(i)}$, e.g. $x^{(i)} = S_j$ if the system is in state S_j at time i. Then the sequence of $x^{(i)}$ from time zero to the end of the walk forms the chain. This sequence of events is a *Markov* chain if the state-to-state transition probabilities are independent of time and history — that is, they depend only on the current state of the system, not on how or when it got there. We define

$$P_{kj} \equiv P(x^{(i+1)} = S_k \leftarrow x^{(i)} = S_j) \tag{1.44}$$

as the probability of the system changing from state S_j to S_k in one time step. Note that it is common practice to list the final state to the left of the initial state. P_{kj} is independent of i for a Markov process. Normalization requires that the rows of $\mathbf{P}$ sum to unity, i.e.

$$\sum_{k=1}^{N} P_{kj} = 1. \tag{1.45}$$

This means that with unit probability the system will be in one of its N allowed states at the next time step. Let the probability that the system is in state S_k at time i be $p_k^{(i)}$. The probability-space density may then be represented as the column vector,

$$\mathbf{p}^{(i)} = \begin{bmatrix} p_1^{(i)} \\ \vdots \\ p_N^{(i)} \end{bmatrix}. \tag{1.46}$$

Normalization of the p's gives $\sum_k^N \mathbf{p}_k^{(i)} = 1$.

At each point in time the system may move from S_j to S_k with probability P_{kj}. The probability distribution thus will evolve as

$$p_k^{(i+1)} = \sum_j P_{kj} p_j^{(i)}, \tag{1.47}$$

which can be written in matrix form as $\mathbf{p}^{(i+1)} = \mathbf{P}\mathbf{p}^{(i)}$. Using this notation, the system evolves from the initial distribution as follows: $\mathbf{p}^{(1)} = \mathbf{P}\mathbf{p}^{(0)}$; and then $\mathbf{p}^{(2)} = \mathbf{P}\mathbf{p}^{(1)} = \mathbf{P}\mathbf{P}\mathbf{p}^{(0)}$. After m steps, $\mathbf{p}^{(m)}$ is related to $\mathbf{p}^{(0)}$ by

$$\mathbf{p}^{(m)} = \mathbf{P}^m \mathbf{p}^{(0)}. \tag{1.48}$$

After a sufficiently long time M, $|\mathbf{p}^{(M+1)} - \mathbf{p}^{(M)}| \to 0$. This implies the existence of an *equilibrium* probability distribution, $\mathbf{p}^*$, defined by

$$\mathbf{p}^* = \mathbf{P}\mathbf{p}^*. \tag{1.49}$$

This equilibrium distribution is a stationary state or fixed point distribution[5] of the transition matrix $\mathbf{P}$.

To illustrate this. consider a simple two state Markov process with the following transition matrix,

$$\mathbf{P} = \begin{pmatrix} 1/4 & 1/8 & 2/3 \\ 3/4 & 5/8 & 0 \\ 0 & 1/4 & 1/3 \end{pmatrix}. \tag{1.50}$$

If one chooses the initial state such that $p_1^{(0)} = 1, p_2^{(0)} = 0, p_3^{(0)} = 0$, then the first three steps are,

$$\begin{bmatrix} 1 \\ 0 \\ 0 \end{bmatrix} \to \begin{bmatrix} 1/4 \\ 3/4 \\ 0 \end{bmatrix} \to \begin{bmatrix} 5/32 \\ 21/32 \\ 6/32 \end{bmatrix}. \tag{1.51}$$

Table 1.2 shows the convergence of the above process in probability space. The steady state fixed point is found by solving the set of equations,

$$\begin{aligned}
\mathbf{p}_1^* &= 1/4\mathbf{p}_1^* + 1/8\mathbf{p}_2^* + 2/3\mathbf{p}_3^* \\
\mathbf{p}_2^* &= 3/4\mathbf{p}_1^* + 5/8\mathbf{p}_2^* \\
\mathbf{p}_3^* &= 1/4\mathbf{p}_2^* + 1/3\mathbf{p}_3^*,
\end{aligned} \tag{1.52}$$

with the constraint that $\mathbf{p}_1^* + \mathbf{p}_2^* + \mathbf{p}_3^* = 1$. This yields the final state of $\mathbf{p}_1^* = 4/15$, $\mathbf{p}_2^* = 8/15$, $\mathbf{p}_3^* = 3/15$.

Table 1.2: Convergence of a three-state Markov process.

Iteration	p_1	p_2	p_3
0	1.00000	0.00000	0.00000
1	0.25000	0.75000	0.00000
2	0.15625	0.65625	0.18750
3	0.24609	0.52734	0.22656
4	0.27848	0.51416	0.20736
5	0.27213	0.53021	0.19766
6	0.26608	0.53548	0.19844
7	0.26575	0.53424	0.20002
8	0.26656	0.53321	0.20023
9	0.26678	0.53318	0.20005
10	0.26671	0.53332	0.19998
11	0.26666	0.53335	0.19999
12	0.26666	0.53334	0.20000
13	0.26667	0.53333	0.20000
$\mathbf{p}^*$	0.26667	0.53333	0.20000

Let us now generalize the above discussion to continuous variables. For concreteness, consider a particle moving in an external potential. Now both space and time will be taken to be continuous. The probability of moving from point $\mathbf{x}$ at time t, to $\mathbf{y}$ at time $t + \Delta t$ is denoted $G(\mathbf{y}, \mathbf{x}; \Delta t)$ — the continuous analog of the matrix P_{kj}. Let the probability density for a particle at $\mathbf{x}$ at time t be $f(\mathbf{x}, t)$, the counterpart of $\mathbf{p}^{(i)}$. Eq. 1.47 and Eq. 1.48 become

$$f(\mathbf{y}, t + \Delta t) = \int f(\mathbf{x}, t) G(\mathbf{y}, \mathbf{x}; \Delta t) d\mathbf{x} \tag{1.53}$$

and

$$f(\mathbf{y}, t + m\Delta t) = \int f(\mathbf{x}, t) G(\mathbf{y}, \mathbf{x}; m\Delta t) d\mathbf{x}, \tag{1.54}$$

with integration now replacing summation.

Again, with some relatively weak conditions on $G(\mathbf{y}, \mathbf{x}; t)$, there exists an equilibrium distribution function $f^*(\mathbf{y})$, which is independent of time, i.e.

$$f^*(\mathbf{y}) = \int f^*(\mathbf{x}) G(\mathbf{y}, \mathbf{x}; t) d\mathbf{x}. \qquad (1.55)$$

Iteration of Eq. 1.53 will therefore produce the continuous distribution $f^*(\mathbf{y})$.

1.3.2   Random walks in state space

In the above discussion we have focused on the evolution of the probability density function along a Markov chain. Let us now return to the evolution of the distribution in the *state space* S_k. A single walker will be in a single state at any given time. If the walker occupies state S_k, then we can write its state-space distribution as $\mathbf{x}^{(i)} = (\delta_{1,k}, \cdots, \delta_{N,k})$, where δ_{ij} is the Kronecker delta. So what does it mean for a Markov chain to converge to an equilibrium density? For a single walker, equilibrium refers to the probability density with which the states are sampled in time, i.e., during the walk states are sampled with probability $\mathbf{p}^*$. Aside from random fluctuations, all averages will be independent of time. A histogram of the walk may be constructed by counting the number of times a walker visits each state and dividing by the total number of counts. This histogram is a map of the equilibrium state distribution.

One of the conditions, alluded to earlier, necessary for the random walk to reach equilibrium is *ergodicity*. A process is *ergodic* if the spatial averages in the limit of an infinite system are equal to the temporal average just discussed. For a process to be ergodic it is necessary (although not sufficient) that all possible states must have a nonzero probability of being visited. For example, if the transition probabilities to a state S_k are all zero, i.e., $P_{kj} = 0$ for all j, the chain cannot be ergodic. Note that it is not always as easy to detect an isolated state as in this simple example.

To understand this concept better, suppose that a single walker visits the points $X^{(0)}, X^{(1)}, \cdots, X^{(m)}$ during the walk. The time average of a function $f(x)$ during this walk is given by,

$$\langle f \rangle_{\{t\}} \equiv \frac{1}{m} \sum_{i=1}^{m} f(X^{(i)}). \tag{1.56}$$

Once equilibrium has been achieved, $\langle f \rangle_{\{t\}}$ is independent of the starting point and time. Now, rather than following a single walker, consider an *ensemble* of walkers $\{X\} = X_1, X_2, \cdots, X_N$, each performing independent random walks. The ensemble, or spatial, average is

$$\langle f \rangle_{\{X\}} \equiv \frac{1}{N} \sum_{k=1}^{N} f(X_k). \tag{1.57}$$

If the ensemble is also drawn from the equilibrium distribution, then the two averages, Eq. 1.56 and Eq. 1.57, are equivalent. In such a situation we may average over time and space in any combination to obtain,

$$\langle f \rangle_{\mathbf{p}^*} = \frac{1}{Nm} \sum_{i=1}^{m} \sum_{k=1}^{N} f(X_k^{(i)}). \tag{1.58}$$

Taking either the limit $m \to \infty$ or the limit $N \to \infty$ then makes this average exact. Depending upon the situation at hand, we may usefully trade off between spatial and temporal averaging.

We have now seen that a stationary distribution, $\mathbf{p}^*$, can arise from a Markov process governed by a regular probability transition matrix $\mathbf{P}$. However, we do not yet have direct control over $\mathbf{p}^*$, as we did in the techniques of Sec. 1.2. Instead, $\mathbf{p}^*$ is a consequence of $\mathbf{P}$. If we wish to generate a particular distribution, we require a method to invert the above procedure to find the appropriate $\mathbf{P}$ for the desired $\mathbf{p}^*$. This is accomplished by the *Metropolis method*.

1.3.3   The Metropolis method

Consider again the discrete N-state system, with equilibrium probabilities $\mathbf{p}^*$. Further, say that the state of maximum probability is S_i, i.e. $p_i^* = \max(\mathbf{p}^*)$. We can then use an acceptance/rejection step similar to that in the rejection method we described earlier. The acceptance probabilities are chosen such that the probability of moving from this most probable state S_i to any other state, S_k, is $A_{ki} = p_k^*/p_i^*$. This choice insures that the relative probability of states S_i and S_k is consistent with the probability density function. Similarly, for the second most probable state S_j, the probability of moving from j to k ($k \neq i$) must once again be $A_{kj} = p_k^*/p_j^*$. One may readily demonstrate that these transition probabilities also obey the relation $A_{kj} \cdot A_{ji} = A_{ki}$, so that the probability of moving from i to k is the same regardless of the path.

The above argument may be continued until the entire triangular matrix A_{ji} with $p_i^* \geq p_j^*$ is constructed. (We will do this for an example to get Eq. 1.61 below.) The remaining matrix elements A_{ij} correspond to moves to higher probability. Because $p_j^*/p_i^* > 1$, this ratio is not a valid probability, and we cannot continue the above argument to construct $\mathbf{A}$. However as we will see below, the correct relative probabilities are achieved by setting all these to unity (including the diagonal elements A_{ii}). Hence, if we ordered the states in ascending probability ($p_1^* \leq p_2^* \leq \cdots$), then $\mathbf{A}$ will have p_j^*/p_i^* as its upper triangle and 1 elsewhere.

Let us now examine how the above choice for $\mathbf{A}$ produces the equilibrium distribution. We know that in equilibrium the ratio of populations in two states must be p_i^*/p_j^*. Let ν_i and ν_j be the current population of states S_i and S_j respectively in a very large ensemble, with $p_i^* > p_j^*$. Then all the ν_j walkers at S_j may move to S_i (since $A_{ij} = 1$). The average number of walkers that can move in the reverse direction, from S_i to S_j, is $\nu_i p_j^*/p_i^* = \nu_i A_{ji}$. The *net* change in population between S_j and S_i is therefore,

$$\delta\nu_j = \nu_i p_j^*/p_i^* - \nu_j. \tag{1.59}$$

If the current population of the ensemble were such that $\nu_i/\nu_j = p_i^*/p_j^*$, then $\delta\nu_j = 0$ and we are at equilibrium. If, on the other hand, $\nu_i/\nu_j > p_i^*/p_j^*$ then $\delta\nu_j > 0$ and the population in S_j increases, driving this inequality toward equality and equilibrium.

Similarly if $\nu_i/\nu_j < p_i^*/p_j^*$ then $\delta\nu_j < 0$ and the population in S_j decreases, again driving the system toward equilibrium. Hence, the above choice of **A** correctly drives the ensemble toward a stable equilibrium distribution.

This discussion can be generalized for continuous variables. Adding an explicit continuous time variable, as well as the spatial variables **x** and **y**, the acceptance probabilities may be written

$$A(\mathbf{y}, \mathbf{x}; \Delta t) = \min\left(\frac{\mathbf{p}^*(\mathbf{y}, t + \Delta t)}{\mathbf{p}^*(\mathbf{x}, t + \Delta t)}, 1\right), \qquad (1.60)$$

to account for both cases in which the ratio of the p's are > 1 and < 1. A simple implementation of the Metropolis algorithm for a continuous one-dimensional case is shown schematically for the first few steps in Fig. 1.6 and is outlined in the following algorithm.

ALGORITHM 1.1 Simple Metropolis sampling. This and the following algorithms in this book are in "schematic FORTRAN." By this we mean that not all details are included, nor is everything necessarily in legitimate code.

```
C
C Initialization:
C Choose the ensemble size, M, and randomly choose the initial
C ensemble positions X.  The function URAN(iseed) generates a uniform
C random number between 0 and 1.
C
      Do i=1,M
          X(i) = URAN(iseed)
      End Do
C
C Loop over moves. Choose the number of steps in the walk, Nstep.
C Then propose a move Y = X + Delta*URAN, where Delta is the step
C size.
C
      fave = 0.
      fsqu = 0.
      Do Istep=1,Nstep
          Do i=1,M
              Y = X(i) + Delta*(URAN(iseed) - 0.5)
C
C Compute the acceptance probability.  The function pstar is the density
C function we wish to sample.
C
              A = min(pstar(Y)/pstar(X(i)),1.)
C
C Accept the move with probability A.
```

```
C
                If (A.ge.URAN(iseed)) Then
                    X(i) = Y
                End If
C
C Sample any desired function f(X) and its square.
C The quantities fave and fsqu will only be averages over pstar if
C equilibrium was previously obtained by iteration of the preceding steps.
C
                fave = fave + f(X(i))
                fsqu = fsqu + f(X(i))**2
            End Do
        End Do
C
C Compute the mean and standard deviation.
C
        fmean = fave/M/Nstep
        sigma = Sqrt( fsqu/M/Nstep - fmean**2 ) / Sqrt( M*Nstep - 1 )
C
C Repeat until the desired statistical accuracy is achieved.
C
        End
```

Several points need to be emphasized regarding this algorithm. First, the walk must be allowed to come to equilibrium before the desired averages may be computed. Methods for judging equilibration will vary with application, but typically one can monitor the running average of a function, observing convergence to a steady state value (within statistical fluctuations). A second point is that if the move $\mathbf{Y}$ is rejected, one must again include the point $\mathbf{X}^{(k)}$ in the distribution, and *not* attempt a second move. A final point is that the distribution is normalized to the total number of sample points M during the walk, not to unity. This normalization leads to division by M when obtaining averages or histograms.

Metropolis sampling is best illustrated by a simple example. Consider our earlier example, the discrete three-state distribution whose transition matrix is given by Eq. 1.50. For simplicity, reorder the states in order of ascending probability, $\mathbf{p}_1^* = 3/15$, $\mathbf{p}_2^* = 4/15$, $\mathbf{p}_3^* = 8/15$. Following the above arguments, the acceptance matrix is,

$$\mathbf{A} = \begin{pmatrix} 1 & 3/4 & 3/8 \\ 1 & 1 & 1/2 \\ 1 & 1 & 1 \end{pmatrix}. \tag{1.61}$$

Figure 1.6: Illustration of the Metropolis random walk. The generation number increases along the horizontal axis and the probability increases along the vertical axis. In (a), at time $t = 0$, the point Y_1 (which has a greater probability p^*) is proposed. In (b), at time $t = 1$, Y_1 has been accepted as X_1 and the point Y_2 (with a lesser p^*) is proposed. Point Y_2 is rejected resulting in $X_2 = X_1$. The walk continues in (c) and (d) to Y_3 (which is accepted) and Y_4.

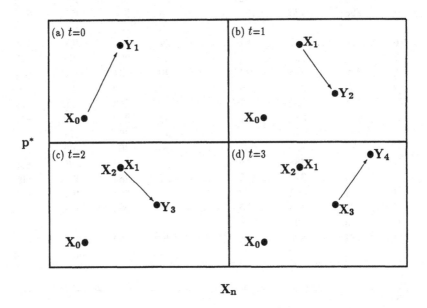

Consider an initial distribution of 1,500 walkers all in state 1, denoted by $X^{(0)} = (1500,0,0)$. At each step in the random walk, let each walker attempt to change to any other state (including itself) with equal probability (i.e., 1/3 for each state). The moves are then accepted with the probabilities of Eq. 1.61. A summary of the first 40 steps is given in Table 1.3. After only six steps the walk has converged to the equilibrium distribution — with only random fluctuations occurring around $X = (300, 400, 800)$.

Table 1.3: Convergence of a three-state Metropolis walk.

Iteration	X_1	X_2	X_3	Iteration	X_1	X_2	X_3
0	1500	0	0	21	331	396	773
1	508	486	506	22	319	396	785
2	344	458	698	23	314	408	778
3	320	428	752	24	318	394	788
4	298	435	767	25	287	415	798
5	289	430	781	26	302	395	803
6	306	393	801	27	280	391	829
7	320	395	785	28	295	382	823
8	302	384	814	29	284	367	849
9	309	391	800	30	282	391	827
10	311	402	787	31	293	406	801
11	315	384	801	32	279	409	812
12	305	404	791	33	294	376	830
13	273	392	835	34	265	409	826
14	297	388	815	35	296	398	806
15	288	393	819	36	264	427	809
16	314	404	782	37	293	401	806
17	305	415	780	38	283	408	809
18	303	391	806	39	294	421	785
19	287	420	793	40	302	372	826
20	302	422	776	$\mathbf{p}^*$	300	400	800

1.3.4   The generalized Metropolis method

In the last example, each walker attempted to move to any state with equal (uniform) probability. Rather than moving based on a uniform density, we can let the probability of attempting a move from state i to j be T_{ji}. We must then accept this move with a probability B_{ji} chosen to produce $\mathbf{p}^*$. What is the appropriate form of $\mathbf{B}$, given $\mathbf{T}$, to produce $\mathbf{p}^*$? Let us apply our previous population balancing argument in this case.

As before, we consider a large ensemble of walkers with ν_i and ν_j as the populations of walkers at i and j, respectively. If $p_j^* < p_i^*$, then the average number of walkers attempting a move from $j \to i$ will be $\nu_j T_{ij}$. These will be accepted with probability B_{ij}. Similarly, from $i \to j$ the average number moving is $\nu_i T_{ji} B_{ji}$. The net increase in population at point j from point i is therefore

$$\delta \nu_j = T_{ji} B_{ji} \nu_i - T_{ij} B_{ij} \nu_j. \tag{1.62}$$

When the walk equilibrates, the average net population change at any point must be zero. Thus, at equilibrium $\delta \nu_j = 0$, and from Eq. 1.62 we require,

$$T_{ji} B_{ji} \nu_i = T_{ij} B_{ij} \nu_j. \tag{1.63}$$

This can be rewritten as

$$\frac{\nu_j}{\nu_i} = \frac{T_{ji}}{T_{ij}} \frac{B_{ji}}{B_{ij}}. \tag{1.64}$$

At equilibrium we know that $\nu_j/\nu_i = p_j^*/p_i^*$; therefore given $\mathbf{T}$ the matrix $\mathbf{B}$ must satisfy

$$\frac{B_{ji}}{B_{ij}} = \frac{p_j^* T_{ij}}{p_i^* T_{ji}}. \tag{1.65}$$

Equation 1.65 is called the *detailed balance* condition — it ensures that in the ensemble the ratio of the populations is the ratio of the $\mathbf{p}^*$'s.

Many different choices of B_{ij} will satisfy detailed balance. A very good choice is

$$B_{ji} = \min(1, q_{ji}), \tag{1.66}$$

where q_{ji} is the ratio

$$q_{ji} = \frac{T_{ij} p_j^*}{T_{ji} p_i^*}. \tag{1.67}$$

By assuming in turn that $T_{ji} p_i^* > T_{ij} p_j^*$ and vice versa, one can see that Eq. 1.66 does indeed satisfy detailed balance. For the continuous case, Eq. 1.66 becomes

$$B(\mathbf{y}, \mathbf{x}; \Delta t) = \min(1, q(\mathbf{y}, \mathbf{x}; \Delta t)) \tag{1.68}$$

where

$$q(\mathbf{y}, \mathbf{x}; \Delta t) = \frac{G(\mathbf{x}, \mathbf{y}; \Delta t)\mathbf{p}^*(\mathbf{y})}{G(\mathbf{y}, \mathbf{x}; \Delta t)\mathbf{p}^*(\mathbf{x})}. \tag{1.69}$$

These choices of B and q constitute the generalized Metropolis method.

ALGORITHM 1.2 Generalized Metropolis sampling.

```
C
C Initialization: Same as Algorithm 1.1.
C
      Do i=1,M
          X(i) = URAN(iseed)
      End Do
C
C Loop over moves. Choose the number of steps in the walk, Nstep.
C Then propose a move Y = X + TRAN, where TRAN is a random number
C chosen from the desired transition probability.
C
      fave = 0.
      fsqu = 0.
      Do Istep=1,Nstep
          Do i=1,M
              Y = X(i) + TRAN(iseed)
C
C Compute the acceptance probability.
C
              q = G(X(i),Y,t)*pstar(Y)/G(Y,X(i),t)/pstar(X(i))
              B = min(q,1.)
C
C Accept the move with probability B.
C
              If (B.ge.URAN(iseed)) Then
                X(i) = Y
              End If
C
C Sample any desired function f(X) and its square.
C
              fave = fave + f(X(i))
              fsqu = fsqu + f(X(i))**2
          End Do
      End Do
C
C Compute the mean and standard deviation.
C
      fmean = fave/M/Nstep
      sigma = Sqrt( fsqu/M/Nstep - fmean**2 ) / Sqrt( M*Nstep - 1 )
C
C Repeat until the desired statistical accuracy is achieved.
C
      End
```

The generalized Metropolis method is of greatest use where a transition function $G(\mathbf{y}, \mathbf{x}; \Delta t)$ is known that generates a probability density function close to $\mathbf{p}^*$.

1.4 Monte Carlo Integration

We now turn our attention to the application of Monte Carlo to the evaluation of definite integrals. Consider the simple one-dimensional definite integral

$$F = \int_a^b f(x)dx. \tag{1.70}$$

By the *mean value theorem* of calculus, this integral is just $(b - a)$ times the average value of f. We can thus approximate F in terms of the finite sums F_N, i.e.

$$F = \lim_{N \to \infty} F_N, \tag{1.71}$$

where

$$F_N = \frac{b - a}{N} \sum_{i=1}^{N} f(X_i). \tag{1.72}$$

The sample points, $\{X_i\}$, should fully cover the domain from a to b. For a finite sum, of course, the value of F_N will fluctuate depending on the choice of $\{X_i\}$.

A simple choice for $\{X_i\}$ is a uniform grid. Such a choice corresponds to the familiar Riemann sum as the grid spacing goes to zero. Grid methods are highly effective for low dimensional integrals or in cases where the integrand can be separated (to a good approximation) into low dimensional parts. However, as the dimensionality d of the space increases, the number of grid points rises as N^d. For fairly low d (e.g. $d \leq 8$) special non-uniform grids and weighting methods, such as Gaussian quadrature, can still be effective. However, for higher dimensions the computational effort rises so rapidly that other alternatives often are preferable. For example, rather than using a fixed grid, one can choose the points $\{\mathbf{X}_i\}$ randomly drawn from a given probability density function by Monte Carlo methods. Sophisticated Monte Carlo integration schemes, often involving Metropolis sampling, have thus found a niche in statistical mechanics and other areas of physics because of the otherwise intractable problems that can be addressed.

1.4.1   Uniform sampling

The Monte Carlo analog of Riemann sums is uniform sampling. Returning to our one-dimensional example from above, a finite sum F_N is taken over the set of N points $\{X_i\}$ sampled uniformly over the interval $[a, b]$. As with all approximate methods, we desire an estimate of our error, $|F - F_N|$. By the central limit theorem, for large N the set of all sums F_N over different $\{X_i\}$ will have a Gaussian density. The standard deviation of this density, σ_N, is the standard deviation of the mean from Eq. 1.30, i.e.,

$$\sigma_N^2 = \frac{\frac{b-a}{N} \sum_{i=1}^{N} f^2(X_i) - F_N^2}{N - 1}. \tag{1.73}$$

As discussed earlier (Sec. 1.1.3), σ_N gives us a measure of the uncertainty of the measurement: the probability that F lies within $F_N \pm \sigma_N$ is ~ 0.68 and between $F_N \pm 2\sigma_N$ is ~ 0.95. However, for a given pseudorandom number generator there exist only a finite number of different pseudorandom numbers, corresponding to the cycle length. Thus the ultimate precision is limited. In particular, Eq. 1.71 and Eq. 1.73 will cease to converge for N greater than the cycle length (because the same pseudorandom numbers are being used over again). This is perhaps not such a serious problem, because the period length can be made quite large by using a better pseudorandom number generator.

To illustrate this simple type of Monte Carlo integration, consider the evaluation of erf(y) (Eq. 1.7). Table 1.4 lists the Monte Carlo estimate and σ_x for several values of y. The first set of results was computed using 10,000 points distributed uniformly between 0 and y. In this example, Monte Carlo integration encounters a serious problem: as y increases, the Monte Carlo estimate of erf(y) and σ looses accuracy significantly. At $y = 100$, erf(y) is in error by 12.72% and σ is roughly 20 times smaller than the actual error. This occurs because for a fixed number of sample

Table 1.4: Monte Carlo estimates of the error function erf(y) with and without importance sampling.

Argument y	Uniform Sampling erf(y)	σ	Importance Sampling erf(y)	σ	Exact erf(y)
0.01	0.0113	0.0011	0.0113	0.0011	0.0113
0.10	0.1125	0.0032	0.1125	0.0032	0.1125
1.00	0.8420	0.0019	0.8427	0.0035	0.8427
2.00	0.9900	0.0053	0.9919	0.0035	0.9953
10.0	1.0297	0.0057	0.9996	0.0033	1.0000
100.0	1.1272	0.0066	1.0019	0.0033	1.0000

points, as one expands the integration interval the density of points drops. Yet the integrand e^{-x^2} remains peaked at $x = 0$. One is wasting effort sampling the tails of the distribution which make a negligible contribution to the sum. It would be better if the majority of sample points could be clustered in the region where the integrand is large, rather than being distributed uniformly. Monte Carlo methods for achieving such clustering are called *importance sampling*.

1.4.2 Importance sampling

One simple method to sample preferentially in regions where the integrand is large is to use composite "random grids". In the above example of the error function, the integrand is Gaussian with a width σ. Hence we could choose N_1 points between 0 and 2σ, and let the remaining N_2 points sample from $2\sigma_g$ to y. Following this idea further, one could break up the interval into many regions, and then optimize the amount of sampling done in each region.

A more general form of importance sampling involves a comparison function. Suppose we have a function, $w(x)$, defined over the domain $[a, b]$, such that $w(x) \approx f(x)$. Further, suppose we can generate pseudorandom variates drawn from the normalized function $p(x)$, where

$$p(x) = \frac{w(x)}{\int\limits_a^b w(x)dx}, \tag{1.74}$$

is a probability density function over $[a, b]$. This function might be generated in a number of ways, but for now assume that the indefinite integral of $p(x)$ is known and is invertible. Then, by the inversion method of Sec. 1.2.2, random numbers may be sampled directly from $p(x)$. These random "grid points" will have greater density where $f(x)$ is large and lesser density where $f(x)$ is small, to the extent that w approximates f. The integral is now obtained by defining the function $g \equiv f/p$, and rewriting Eq. 1.70 as

$$F = \int\limits_a^b g(x)p(x)dx \approx \frac{1}{M}\sum_{i=1}^M g(X_i). \tag{1.75}$$

The last equality in Eq. 1.75 (in the limit of $M \to \infty$) follows since the $\{X_i\}$ are a random sequence of numbers generated from $p(x)$.

Writing the integral F in the form of Eq. 1.75 is advantageous for two reasons. First, fluctuations in g are greatly reduced over those originally present in f because the summand is now $g(X) = f(X)/p(X) = [f(X)/w(X)]\int w(x)dx$. This quantity is almost a constant independent of X. To see this, consider what happens if $w(x)$ were exactly $f(x)$: each sample point $g(X)$ is now exactly F, and so there is no variance. Choosing $w(X)$ close to $f(X)$ therefore minimizes fluctuations in $g(X)$ and hence in the Monte Carlo estimate of F. The second major advantage of importance sampling is that it allows the domain $[a, b]$ to become infinite. Returning, for example, to the evaluation of $\operatorname{erf}(y)$, the second column of Table 1.4 shows the effects of importance sampling using the exponential density, e^{-x} for w. Unlike in the earlier case of uniform sampling, with importance sampling both the integral and σ are estimated accurately for all y.

1.4.3 Expected values using importance sampling

In quantum mechanics, as in statistical mechanics and in other areas as well, one often requires the expectation value of a function $h(\mathbf{x})$, with respect to a function $f(\mathbf{x})$, i.e.,

$$\langle h \rangle_f \equiv \frac{\int_{-\infty}^{\infty} h(\mathbf{x}) f(\mathbf{x}) d\mathbf{x}}{\int_{-\infty}^{\infty} f(\mathbf{x}) d\mathbf{x}}, \qquad (1.76)$$

rather than integrals of the form of Eq. 1.70. The above expression is directly related Eqs. 1.74 and 1.75, with f and h replacing w and g respectively, and in arbitrary dimensions. Hence, such expectation values may be evaluated directly by the use of importance sampling. To do so we sample $\{\mathbf{X}\}$ from the probability density function $f(\mathbf{x})/\int f(\mathbf{x}')d\mathbf{x}'$ and then average $h(\mathbf{X})$ over this distribution. The integral in the denominator of Eq. 1.76 does not need to be explicitly computed since it is incorporated into the probability density function one samples. Importance sampling is therefore ideally suited for the computation of expectation values. Moreover, the Metropolis method discussed previously allows virtually any function $f(\mathbf{x})$ to be sampled. It is important to note, however, that if just the integral of $h(\mathbf{x})$ were desired, then this approach requires the integral of $f(\mathbf{x})$ to be evaluated separately.

Exercises

1. Derive formulas for the mean, second, and all higher central moments of the uniform and Gaussian distributions. For the Gaussian density function the following integrals will be useful:

$$\int_0^{\infty} x^{2n} e^{-ax^2} dx = \frac{1 \cdot 3 \cdot 5 \cdots (2n-1)}{2^{n+1}} \sqrt{\frac{\pi}{a^{2n+1}}}, \qquad (1.77)$$

$$\int_0^{\infty} x^{2n+1} e^{-ax^2} dx = \frac{n!}{2a^{n+1}}. \qquad (1.78)$$

2. Show that $\text{Cov}(x_1, x_2)$ of the bivariate Gaussian distribution, Eq. 1.11, is $\rho\sigma_1\sigma_2$.

3. Write a computer program for a linear congruential generator. Characterize the resulting distribution using the tests in Sec. 1.2.1. Use Monte Carlo integration to compute the lowest eight moments of the uniform and Gaussian distributions. Compare these with the analytical results of Exercise 1.

4. Use both the inversion and rejection methods to generate pseudorandom numbers between zero and π distributed according to $\sin x$.

5. Write a program that uses the Metropolis method to compute the moments of the bivariate Gaussian distribution, Eq. 1.11, with $\sigma_1 = \frac{1}{2}$, $\sigma_2 = 2$ and $\rho = 2$.

6. Compute the value of π using Monte Carlo integration by finding the area of a circle of unit radius. (Note that this area can be computed as a fraction of the area of a square with a side length of 2 by the rejection method.)

7. Compute by Monte Carlo the exponential integral,[2]

$$\alpha_n(z) = \int_1^\infty t^n e^{-zt} dt \qquad (n = 1, 2, ...; \ z > 0), \tag{1.79}$$

using two different importance sampling methods.

Suggestions for Further Reading

1. H. Gould and J. Tobochnik, *An Introduction to Computer Simulation Methods, Part 2* (Addison-Wesley, 1988).

2. R. Y. Rubinstein, *Simulation and the Monte Carlo Method* (Wiley, New York, 1981).

3. D. W. Heermann, *Computational Simulation Methods in Theoretical Physics* (Springer-Verlag, Berlin, 1986).

4. M. H. Kalos and P. A. Whitlock, *Monte Carlo Methods Volume 1: Basics* (Wiley, New York, 1986).

References

1. There are many good introductory statistics texts, such as W. Feller, *An Introduction to Probability Theory and its Applications* (Wiley, New York, 1968).

2. See, for example, M. Abramowitz and I. Stegun, eds. *Handbook of Mathematical Functions* (Dover, New York, 1972).

3. D. E. Knuth, *Seminumerical Algorithms*, 2nd ed., Volume 2 of *The Art of Computer Programming* (Addison-Wesley, Reading, 1981).

4. W. H. Press, B. P. Flannery, S. A. Teukolsky and W. T. Vetterling, *Numerical Recipes* (Cambridge University Press, Cambridge, 1986).

5. J. M. Hammersley and D. C. Handscomb, *Monte Carlo Methods* (Chapman and Hall, London, 1964).

6. N. Metropolis, A. W. Rosenbluth, M. N. Rosenbluth, A. M. Teller and E. Teller, "Equations of State Calculations by Fast Computing Machines," *Journal of Chemical Physics* **21**, 1087-92 (1953).

Chapter 2

Variational Methods

In this chapter, and throughout the remainder of this book, we will be primarily concerned with solving the non-relativistic, time-independent electronic Schrödinger equation,

$$\mathcal{H}\Phi = E\Phi, \tag{2.1}$$

where $\mathcal{H}$ is the molecular Hamiltonian operator. $\mathcal{H}$ is composed of the kinetic and potential energy operators, i.e., $\mathcal{H} = \mathcal{T} + \mathcal{V}$. For N_e electrons and N_n nuclei with nuclear charges Z_A, these operators are,

$$\mathcal{T} = -\sum_{i=1}^{N_e} \frac{1}{2}\nabla_i^2 \tag{2.2}$$

and

$$\mathcal{V} = -\sum_{i=1}^{N_e}\sum_{A=1}^{N_n} \frac{Z_A}{r_{iA}} + \sum_{i=1}^{N_e}\sum_{j>i}^{N_e} \frac{1}{r_{ij}} + \sum_{A=1}^{N_n}\sum_{B>A}^{N_n} \frac{Z_A Z_B}{R_{AB}}, \tag{2.3}$$

where we have used atomic units, wherein $\hbar$, the electron mass, and the electron charge all are equal to unity. Conversions from atomic units are given in Appendix A.

The variational method is one of the most productive approaches for finding approximate solutions of the electronic Schrödinger equation. According to the variational principle, the expectation value of the energy of a trial wave function Ψ, given by

$$E[\Psi] = \frac{\int \Psi^* \mathcal{H} \Psi \, d\mathbf{x}}{\int \Psi^* \Psi \, d\mathbf{x}}, \tag{2.4}$$

will be a *minimum* for the exact ground state wave function. (For bound electronic states, Ψ may be assumed to be real, so henceforth we assume that Ψ is equal to its complex conjugate Ψ^*.) The functional $E[\Psi]$ thus provides an *upper* bound to the exact ground state energy — the best choice of Ψ may be obtained by minimizing $E[\Psi]$ with respect to the variation of its parameters.

The variational principle does not specify the form of Ψ. To judge which among a set of proposed forms of Ψ is best, the integration indicated by Eq. 2.4 must be accomplished. The ability to integrate Ψ is so important that *ab initio* electronic structure methods have as a primary requirement, aside from antisymmetry, the integrability of quantities such as Eq. 2.4. Consequently, the chosen forms of trial functions, typically sums of Slater determinants formed from one-electron Gaussian-type functions, are only fair representations of the actual wave function, and are only very slowly convergent to the exact wave function. What other choice of Ψ is there? General forms of high accuracy such as the Hylleraas functions[1, 2] have been restricted to atoms with a maximum of four electrons and to molecules with a maximum of two electrons because of integration difficulties. Fully numerical solutions obtained using grid methods are possible for few particles or high symmetry, as exemplified by numerical Hartree-Fock methods for atoms and diatomics.

This chapter focuses on the use of Monte Carlo methods to perform the necessary integration and to minimize $E[\Psi]$ for *arbitrary* forms of Ψ. This approach is referred to as variational Monte Carlo (VMC). Historically, VMC has been little used because Monte Carlo methods require relatively large amounts of computer time. In recent years, however, advances in both algorithms and in the speed of computers have allowed VMC to advance greatly, being perhaps most hampered now by the lack of knowledge about how to construct Ψs of high accuracy.

We begin our discussion with a review of the variational method. We then present Monte Carlo methods for determining $E[\Psi]$ and for optimizing a given trial function form. The broader issue of the choice of the form of Ψ is germane to all the Monte Carlo methods we will discuss in this book, and is deferred to Ch. 5.

2.1 Review of the Variational Method

By providing a bound on the ground state energy, the variational method is a powerful tool for approximately solving the Schrödinger equation. A form for Ψ is chosen that contains a set of variational parameters, $\{\alpha\} \equiv \{\alpha_1, \cdots, \alpha_n\}$. The ultimate accuracy of the method depends solely on the functional form and the variational parameters. Minimizing $E[\Psi]$ with respect to $\{\alpha\}$ involves either analytic differentiation of $E[\Psi]$, or a numerical method of minimization. If practical, we also may wish to compute the second derivatives, both to insure that a true minimum has been obtained and to improve the convergence of the optimization. Consider H_2 as an example: label the electrons as 1 and 2 and the two protons as A and B. Various accurate forms for the wave function are discussed at greater length in Ch. 5. Consider the simple form

$$\Psi(H_2) = (e^{-\alpha r_{1A}} + e^{-\alpha r_{1B}})(e^{-\alpha r_{2A}} + e^{-\alpha r_{2B}}), \tag{2.5}$$

where r_{1A}, r_{1B}, r_{2A} and r_{2B} are the various electron-nucleus distances, and α is the sole variational parameter. Note that Eq. 2.5 is not normalized. Nevertheless, this will not be a problem for us since, as pointed out in Sec. 1.4.3, Monte Carlo evaluation of expectation values does not require explicit knowledge of the normalization.

Minimization of $E[\Psi]$ provides an *upper* bound to the true energy. There are, however, also methods of finding rigorous *lower* bounds to the energy. One such lower bound[3] is the rightmost inequality in the following relationship

$$E[\Psi] + \sigma^2[\Psi] \geq E[\Psi] \geq E_0 \geq E[\Psi] - \sigma^2[\Psi], \tag{2.6}$$

where E_0 is the exact ground state energy and $\sigma^2[\Psi]$ is defined as

$$\sigma^2[\Psi] = \int (\mathcal{H}\Psi)^2 d\mathbf{x} - E^2. \tag{2.7}$$

This latter quantity is the variance of the local energy about the mean. Unfortunately, it has been found that expressions for lower bounds are significantly less strongly bounding than the variational upper bound and, in most cases, they are impractical to evaluate by any method other than Monte Carlo.

Nevertheless, one important consequence of Eq. 2.6 is that one can minimize $\sigma^2[\Psi]$ rather than $E[\Psi]$. This has many advantages when Monte Carlo methods are used to perform the minimization — one of which is that the exact minimum value of $\sigma^2[\Psi]$ is zero, whereas the minimum of $E[\Psi]$, is unknown.

Despite its widespread applicability and simplicity, the variational method is in no way the panacea. For example, perturbation theory methods (e.g. Møller-Plesset and coupled cluster theory) in which the energy can lie either above or below the exact value, may provide greater accuracy. The variational method has gained popularity largely due to the cancellation of errors which occurs when computing chemically important energy differences. Because errors in the variational energies are always positive, these errors must cancel to some extent when an energy difference is computed. For a non-variational method these individual errors may be in either direction and thus may combine instead of canceling. Typically, however, cancellation of errors occurs in both variation and perturbation methods.

2.2 Monte Carlo Evaluation of Expectation Values

To afford the possibility of using compact, high accuracy forms for Ψ — ones that are not analytically integrable — one can integrate numerically. Specifically, for the most general forms of Ψ, in which Ψ is not necessarily constructed from an orbital expansion, Monte Carlo integration is often the method of choice. Here we apply the integration methods of Ch. 1 to the evaluation of $E[\Psi]$.

2.2.1 Simple Metropolis sampling

The straightforward application of uniform sampling to integrate $E[\Psi]$ is doomed to failure for the same reason that grid methods fail: inefficiency. An efficient method generates points based on the integrand. To see how to do this, let us recast Eq. 2.4 as

$$E[\Psi] = \frac{\int \Psi^2(\mathbf{x}) E_L(\mathbf{x}) d\mathbf{x}}{\int \Psi^2(\mathbf{x}) d\mathbf{x}} \equiv \langle E_L \rangle_{\Psi^2} , \tag{2.8}$$

where E_L is the "local energy" defined as $E_L \equiv \mathcal{H}\Psi(\mathbf{x})/\Psi(\mathbf{x})$. To illustrate, in Fig. 2.1 we show the local energy surface for the simple form of Ψ given in Eq. 2.5. This rewriting of Eq. 2.4 in terms of the local energy serves two purposes. First, Eq. 2.8 is now in the form of a weighted average rather than an operator expectation value. The weight here is the normalized probability density function of the electrons $\Psi^2(\mathbf{x})/\int \Psi^2 d\mathbf{x}$. Second, E_L has the property that it is a constant for an eigenfunction of $\mathcal{H}$: since $\mathcal{H}\Phi_k = E_k \Phi_k$, we have $E_L[\Phi_k] = E_k$. This property is significant because this means that Eq. 2.8 can give E_k with zero variance. In practice, Ψ will rarely be an eigenfunction. Nevertheless, the more accurate Ψ becomes, the less variance E_L will have.

If the Monte Carlo sample points, $\{\mathbf{X}\}$, are drawn from the distribution Ψ^2, then

Figure 2.1: Local energy of simple Ψ (Eq. 2.5) for H_2 as a function of electron 1. Electron 2 is located under the large positive peak and the nuclei are located at the two negative peaks.

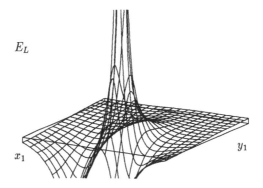

the Monte Carlo estimate of the energy is

$$E[\Psi] = \langle E_L \rangle_{\Psi^2} = \lim_{M \to \infty} \frac{1}{M} \sum_{i=1}^{M} E_L(\mathbf{X}_i). \qquad (2.9)$$

Assuming uncorrelated sampling, the variance of the mean value is given by

$$\sigma^2(E[\Psi]) = \frac{\langle E_L^2 \rangle_{\Psi^2} - \langle E_L \rangle_{\Psi^2}^2}{M - 1}. \qquad (2.10)$$

Recall that the bracket notation $\langle \rangle_f$, introduced in Ch. 1, denotes the average with respect to the distribution f. In general, Ψ^2 will not be of a form for which the distribution can be sampled directly. In such situations one can use Metropolis sampling (cf. Ch. 1) to sample from Ψ^2. A simple implementation follows.

ALGORITHM 2.1 Simple variational Monte Carlo.

```
C
C Initialize run: select the ensemble size M and choose randomly the
C initial ensemble positions X. N is the number of electrons. The function
C URAN(iseed) returns a uniform random number between 0 and 1.
C
      Do k=1,M
         Do i=1,N
            Do j=1,3
               X(i,j,k) = URAN(iseed)
            End Do
         End Do
      End Do
C
C Loop over moves. Choose the number of steps in the walk, Nstep.
C Then propose a move Y = X + Delta*URAN, where Delta is the step
C size. Here we move one electron at a time. Subroutine PSI computes
C the initial value of Psi, then NEWPSI changes the coordinates of
C a single electron and updates Psi. ELOCAL computes the local energy.
C
      Eave = 0.
      Esqu = 0.
      Accept = 0.
      Do Istep=1,Nstep
         Do k=1,M
            Call PSI(PsiX,X(k))
            Do i=1,N
C
C Compute wave function Psi
C
               Do j=1,3
                  Y(j) = X(i,j,k) + Delta*(URAN(iseed) - 0.5)
               End Do
C
C Compute acceptance probability.
C
               Call NEWPSI(PsiY,Y,i)
               A = min(PsiY**2/PsiX**2,1.)
C
C Accept move with probability A.
C
               If (A.ge.URAN(iseed)) Then
C
C Copy Y back into X.
                  Call COPY(X(k),Y)
                  Accept = Accept + 1/N
               End If
            End Do
C
C Sample local energy
C
            Ex = ELOCAL(X(k))
            Eave = Eave + Ex
            Esqu = Esqu + Ex**2
         End Do
      End Do
C
C Compute acceptance ratio, mean, and standard deviation.
C
      Aratio = Accept/M/Nstep
```

```
        Emean = Eave/M/Nstep
        Esigma = Sqrt( Esqu/M/Nstep - Emean**2 ) / Sqrt( M*Nstep - 1 )
C
C Repeat until desired statistical accuracy is achieved.
C
        End
```

An advantage of the Metropolis algorithm is that it only requires evaluating Ψ for the proposed move; the unknown normalization $\int \Psi^2 d\mathbf{x}$ is not required. Initially, one must wait for equilibrium (i.e. for convergence to the distribution Ψ^2) before computing any expectation values. Equilibration may be judged by looking for a systematic trend in $\langle E_L \rangle$ over the course of the walk. Other quantities of interest also can be sampled at the same time. In choosing the step size, one wishes to maximize the actual *accepted* step size. Attempting too large a step will result in a small acceptance to rejection ratio, and so actual movement will be small. Not attempting large enough moves clearly also will restrict the actual movement. This step size should therefore be optimized empirically based on the observed behavior of the sampling algorithm.

In the above algorithm we have moved and accepted or rejected one electron at a time. One also could have moved all the electrons at once, and then accepted or rejected the move as a whole. Moving one electron at a time is often more efficient, especially with the wave function forms typically used in which the one-electron contribution to Ψ and E_L can be evaluated efficiently.

2.2.2   Importance sampling: Fokker-Planck formalism

The greatest shortcoming of the above algorithm is that the random displacement is proposed without any knowledge of Ψ^2, leading to a higher rejection rate than necessary. In this section we introduce a different approach, which does not require rejection, and includes importance sampling based on the Fokker-Planck equation. This formalism, however. leads to a step-size bias in the distribution and energies.

Hence, in Sec. 2.2.3 we will combine the best aspects of both the Metropolis and Fokker-Planck methods to achieve an unbiased Monte Carlo method with importance sampling.

Consider a diffusion process characterized by a time-dependent density $f(\mathbf{x}, t)$. Simple isotropic diffusion processes obey the following Fokker-Planck equation,

$$\frac{\partial f}{\partial t} = \sum_i D \frac{\partial}{\partial \mathbf{x}_i} \left(\frac{\partial}{\partial \mathbf{x}_i} - F_i(\mathbf{x}) \right) f, \tag{2.11}$$

where D is the diffusion constant and F_i is the i-th component of a drift velocity $\mathbf{F}$ caused by an external potential. As before, we wish to converge to the stationary density $f = \Psi^2 / \int \Psi^2 d\mathbf{x}$. An unchanging state, for which $\partial f / \partial t = 0$, may be obtained by setting the left-hand side of Eq. 2.11 to zero, namely

$$\sum_i D \left(\frac{\partial^2 f}{\partial \mathbf{x}_i^2} - \frac{\partial}{\partial \mathbf{x}_i} (F_i f) \right) = 0. \tag{2.12}$$

Equation 2.12 can most readily be satisfied if each term of the sum vanishes, yielding

$$\frac{\partial^2 f}{\partial \mathbf{x}_i^2} = f \frac{\partial}{\partial \mathbf{x}_i} F_i + F_i \frac{\partial}{\partial \mathbf{x}_i} f. \tag{2.13}$$

The drift velocity $\mathbf{F}$ therefore must be of the form $F_i = g(f) \partial f / \partial \mathbf{x}_i$ in order to obtain a second derivative of f on the right-hand side. If we substitute this form of $\mathbf{F}$ into Eq. 2.13 we find

$$\frac{\partial^2 f}{\partial \mathbf{x}_i^2} = f \frac{\partial g}{\partial f} \left(\frac{\partial f}{\partial \mathbf{x}_i} \right)^2 + f g \frac{\partial^2 f}{\partial \mathbf{x}_i^2} + g \left(\frac{\partial f}{\partial \mathbf{x}_i} \right)^2. \tag{2.14}$$

Cancellation of the second derivative terms requires that $g = 1/f$. This choice also leads to cancellation of the first derivative terms. Therefore the stationary density $f = \Psi^2 / \int \Psi^2 d\mathbf{x}$ results from choosing the drift vector to be

$$\mathbf{F} = \frac{1}{f} \nabla f = 2 \frac{1}{\Psi} \nabla \Psi. \tag{2.15}$$

Clearly this drift causes the move to be biased by Ψ. This biased diffusion process incorporates importance sampling.

We now have a diffusion equation which gives the desired distribution, but how do we implement it using Monte Carlo? In statistical mechanics, Fokker-Planck trajectories are generated by means of a Langevin equation. The Langevin equation corresponding to Eq. 2.11 is

$$\frac{\partial \mathbf{x}(t)}{\partial t} = D\mathbf{F}(\mathbf{x}(t)) + \eta. \tag{2.16}$$

Here η is a randomly fluctuating force which is distributed according to a multidimensional Gaussian with a mean of zero and a variance of $2D$. The Langevin equation arises from the theory of Brownian motion, which is closely related to diffusion. By integrating it over a short time interval, δt, we obtain a discretized form suitable for Monte Carlo simulation, which moves the particle from point $\mathbf{x}$ to $\mathbf{y}$ according to

$$\mathbf{y} = \mathbf{x} + D\mathbf{F}(\mathbf{x})\delta t + \chi. \tag{2.17}$$

In this equation χ is a Gaussian random variable with a mean value of zero and a variance of $2D\delta t$. By using the discretized form, Eq. 2.17, rather than the continuous form, Eq. 2.16, we have introduced a bias into the dynamics for any $\delta t > 0$. The distribution of trajectories, and therefore the measured energy, will deviate increasingly from the exact as δt increases. If the resulting bias is too large, one may take several separate estimates of E for different values of δt, and extrapolate the results to $\delta t = 0$. However, this error can be corrected by a Metropolis acceptance-rejection step, as we proceed to do in the next section.

2.2.3   Importance sampling: Metropolis formalism

In Sec. 2.2.1 we discussed the use of Metropolis sampling to obtain the distribution Ψ^2. Each step in that random walk was isotropic. In Sec. 2.2.2 we obtained a dynamical equation (Eq. 2.17) for generating guided trajectories containing a vector force, $\mathbf{F} = 2\Psi^{-1}\nabla\Psi$. Equation 2.17 directs the walkers toward regions of large Ψ^2, but suffers by being exact only at $\delta t = 0$. We will now show that by combining Eq. 2.17 with a Metropolis rejection step, the time step bias can be eliminated.

From the generalized Metropolis method, presented in Sec. 1.3.4, recall that the transition probability $P(\mathbf{y}, \mathbf{x})$ is composed of the marginal transition probability $G(\mathbf{y}, \mathbf{x}; \delta t)$ multiplied by $\Psi^2(\mathbf{x})$, the probability of finding a walker at $\mathbf{x}$. We can use this approach, and choose $G(\mathbf{y}, \mathbf{x}; \delta t)$ corresponding to the Fokker-Planck equation, accepting the move with the generalized probability

$$A(\mathbf{y}, \mathbf{x}) = \min(1, q(\mathbf{y}, \mathbf{x})), \tag{2.18}$$

where $q(\mathbf{y}, \mathbf{x})$ is the ratio

$$q(\mathbf{y}, \mathbf{x}) = \frac{G(\mathbf{x}, \mathbf{y}; \delta t)\Psi^2(\mathbf{y})}{G(\mathbf{y}, \mathbf{x}; \delta t)\Psi^2(\mathbf{x})}. \tag{2.19}$$

We now focus on the form of $G(\mathbf{y}, \mathbf{x}; \delta t)$ that corresponds to the Fokker-Planck equation.

As will be shown in Ch. 3, $G(\mathbf{y}, \mathbf{x}; \delta t)$ must be a solution of the Fokker-Planck equation with the added condition that $G(\mathbf{y}, \mathbf{x}; \delta t = 0) = \delta(\mathbf{x} - \mathbf{y})$. To solve the Fokker-Planck equation, it is convenient to rewrite it in the form,

$$\frac{\partial f}{\partial t} = \mathcal{L}f , \tag{2.20}$$

where $\mathcal{L} = D\nabla \cdot (\nabla - \mathbf{F})$. Then $G(\mathbf{y}, \mathbf{x}; \delta t)$ is the spatial resolution of the operator $e^{-\mathcal{L}\delta t}$. This can be confirmed by substitution into Eq. 2.20 (see Exercise 5). In operator form, $G(\mathbf{y}, \mathbf{x}; \delta t)$ is given by

$$G(\mathbf{y}, \mathbf{x}; \delta t) = \exp[D\delta t(\nabla^2 - \nabla \cdot \mathbf{F} - \mathbf{F} \cdot \nabla)]. \tag{2.21}$$

If we now assume that the force $\mathbf{F}$ remains essentially constant between $\mathbf{x}$ and $\mathbf{y}$, we can integrate Eq. 2.21 over a small time interval (which we will still call δt). This makes Eq. 2.21 a function of **x** and **y**, which when normalized becomes

$$G(\mathbf{y}, \mathbf{x}; \delta t) = (4\pi D\delta t)^{-3N/2} \exp[-(\mathbf{y} - \mathbf{x} - D\delta t \mathbf{F}(\mathbf{x}))^2 / 4D\delta t]. \qquad (2.22)$$

Note that the move prescribed by Eq. 2.17 is consistent with Eq. 2.22, which is a Gaussian with variance $2D\delta t$ whose center is drifting due to a vector force field. Moreover, Eq. 2.22 is a solution to the generalized diffusion equation, Eq. 2.11.

The function $G(\mathbf{y}, \mathbf{x}; \delta t)$ gives the probability of a walker moving from **x** to **y**. Therefore the total density at point **y**, i.e. the function $f(\mathbf{y})$, is given by the integral over all space of the transition probabilities multiplied by f at each point, i.e.

$$f(\mathbf{y}, t + \delta t) = \int G(\mathbf{y}, \mathbf{x}; \delta t) f(\mathbf{x}, t) d\mathbf{x}. \qquad (2.23)$$

Repeated iterations of Eq. 2.23 by means of Eqs. 2.17 and 2.18 will therefore produce $f(\mathbf{y}, t \to \infty) = \Psi^2$.

We now give an algorithm for the above importance-sampled variational Monte Carlo algorithm:

ALGORITHM 2.2 Importance sampled variational Monte Carlo.

```
C
C Initialize run: choose the ensemble from Algorithm 2.1 and the number of
C steps in the walk, Nstep.  Then propose a move Y = X + D*DeltaT*F + GRAN.
C Here, as earlier, we move one electron at a time. The functions and
C subroutines are as in Algorithm 2.1 except that PSI and NEWPSI now return
C both the trial function Psi and the force F.
C GRAN returns a Gaussian random number with variance of 2*D*DeltaT.
C
      Eave = 0.
      Esqu = 0.
      Accept = 0.
      Do Istep=1,Nstep
          Do k=1,M
              Call PSI(PsiX,FX,X(k))
              Do i=1,N
C
C Compute wave function Psi
C
              Do j=1,3
                  Y(j) = X(i,j,k) + D*DeltaT*F(j) + GRAN(iseed)
              End Do
```

```
C
C Compute acceptance probability.
C
                  Call NEWPSI(PsiY,FY,Y,i)
C
C To compute the Metropolis acceptance probability first compute the log of the
C ratio of the forward and reverse moves. (This is a useful formula for future
C reference.)
C
                  Q = 0
                  Do j=1,3
                       Q = Q + 1/2*(FX(j) + FY(j))*[D*DeltaT/2*(FX(j) - FY(j))
                                                  - (Y(j) - X(i,j,k)]
                  End Do
                  A = min( 1 , PsiY**2/PsiX**2*EXP(Q) )
C
C Accept the move with probability A and proceed as above.
C
                  If (A > URAN(iseed)) Then
                       Call COPY(X(k),Y)
                       Accept = Accept + 1/N
                  End If
              End Do
              Ex = ELOCAL(X(k))
              Eave = Eave + Ex
              Esqu = Esqu + Ex**2
          End Do
      End Do
      Aratio = Accept/M/Nstep
      Emean = Eave/M/Nstep
      Esigma = Sqrt( Esqu/M/Nstep - Emean**2 ) / Sqrt( M*Nstep - 1 )
      End
```

This algorithm is far from optimal. From the physics alone, improvements can be made by paying attention to singularities, such as in the potential energy where two particles meet, and in the force $\mathbf{F} = 2\nabla\Psi/\Psi$ where Ψ goes to zero. Some of these improvements can be derived from the proper choice of Ψ (see Ch. 5) and some by a more judicious choice of G. In Table 2.1 we show representative energies for He through Ne computed with an algorithm similar to Algorithm 2.2.

2.2.4 Electronic properties

Some properties, such as atomization energies, ionization potentials, and electron affinities, are simply obtained as a difference of energies. Many other properties can be measured as well. Local properties, which are a function of the electronic and nuclear coordinates of the trial function, can be most easily computed. For example,

Table 2.1: Hartree-Fock, VMC, and estimated exact total energies for He through Ne.

System	Hartree-Fock[5]	VMC Energy[4]	Exact[6]	% CE[a]
He	−2.8617	−2.9036(1)	−2.9037	100
Li	−7.4327	−7.4768(3)	−7.4781	97
Be	−14.5730	−14.6370(6)	−14.6673	68
B	−24.5291	−24.6156(6)	−24.6539	69
C	−37.6886	−37.9017(7)	−37.8451	72
N	−54.4009	−54.5456(6)	−54.5895	77
O	−74.8094	−75.0146(7)	−75.0673	80
F	−99.4093	−99.6736(7)	−99.7313	82
Ne	−128.5471	−128.8796(6)	−128.9370	85

(a) % CE (correlation energy) is the difference between the Hartree-Fock and VMC energies relative to the difference between the Hartree-Fock and exact energies.

the dipole moments, μ_x, μ_y and μ_z, can be obtained by summing over the coordinates of each electron, i.e.

$$\mu_x = \sum_{i=1}^{N} \frac{\int \Psi x_i \Psi d\mathbf{x}}{\int \Psi^2 d\mathbf{x}} \tag{2.24}$$

where x_i is the x component of electron i. In Table 2.2 we list VMC-computed second moments of H_2.

2.2.5   Evaluation of the variance

Associated with every Monte Carlo estimate is a variance σ^2. A standard expression for the variance of the mean is

$$\hat{\sigma}^2 = \langle A^2 \rangle - \langle A \rangle^2. \tag{2.25}$$

This expression assumes that successive values of the random quantity, A, are statistically independent. A more accurate expression is

$$\sigma^2 = \hat{\sigma}^2 + \frac{2}{N-1} \sum_{i>j}^{N} \text{Cov}(A_i, A_j), \tag{2.26}$$

Table 2.2: Second moments of H_2. The electric quadrupole moment, Q, is derived from the other expectation values.

Moment	Hartree-Fock[8]	VMC[7]	Exact[9]
$\frac{1}{2}(\langle x^2 \rangle + \langle y^2 \rangle)$	0.7768	0.7715(9)	0.7617
$\langle z^2 \rangle$	1.020	1.078(2)	1.023
$\langle r^2 \rangle$	2.574	2.621(3)	2.546
Q	0.66	0.49(1)	0.61

where for present purposes we have assumed that all the A_i are sampled from the same distribution. A more illustrative way to write Eq. 2.26 is in terms of the correlation coefficient,

$$\rho_k \equiv \frac{1}{\hat{\sigma}^2(N-k)} \sum_{i=1}^{N-k} (A_i - \bar{A})(A_{i+k} - \bar{A}) \tag{2.27}$$

which is a measure of the correlation between points which are k steps apart in the walk. Note that $0 \leq \rho_k \leq 1$. Rewriting Eq. 2.26 in terms of ρ_k yields,

$$\sigma^2 = \hat{\sigma}^2 \left(1 + 2 \sum_{k=1}^{N} \rho_k \right). \tag{2.28}$$

Quite often ρ_k decays exponentially with k, and so there exists a value K_0 such that $\rho_k \approx 0$ for all $k \geq K_0$. The value K_0 is a *correlation length*. Hence the effect of such correlation on Eq. 2.28 can be estimated by summing ρ_k from 1 to K_0. Note that neglect of this serial correlation will always cause the value of our simple estimate, $\hat{\sigma}^2$, to be less than the true σ^2.

In the random walks used here, K_0 strongly depends upon the time step δt. Figure 2.2 shows the dependence of ρ_k on k as a function of step size for a simple Metropolis walk. Clearly, it is preferable to use the largest time step possible to minimize serial correlation. This choice must be balanced in rejection methods with the necessity of using a smaller time step to avoid excessive rejection.

Other methods for estimating σ^2 are often preferred over Eq. 2.28 because of the

Figure 2.2: Typical dependence of the correlation coefficient ρ_k on step size and k in a VMC simulation.

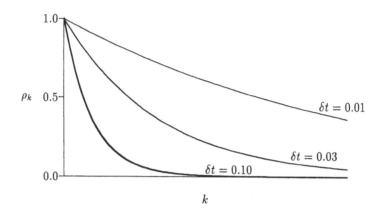

implied double sum. The simplest is the method of data blocking. Consider grouping the data as follows,

$$\langle A \rangle = \frac{1}{n} \sum_{i=1}^{n} \left(\frac{1}{m} \sum_{j=i\times m+1}^{(i+1)m} A_j \right) \equiv \frac{1}{n} \sum_{i=1}^{n} \langle A \rangle_i. \tag{2.29}$$

The quantity $\langle A \rangle_i$ is the block average of all data between $(i \times m) + 1$ and $(i+1)m$, where m is the block size. Figure 2.3 shows schematically the effect of blocking on the local energy. If $m > K_0$, the quantities A_i and A_{i+m} will be essentially uncorrelated, and $\hat{\sigma}^2$ will be very close to σ^2. It is only necessary to estimate empirically the value of K_0 in order to implement blocking. A possible pitfall of this method is that as K_0 increases, one may be unable to compute a sufficient number of blocks to accurately estimate σ^2.

Another method for estimating the true σ^2 is to perform independent walks. For example, for n independent walks started sufficiently far from each other, $\hat{\sigma}^2$ will

Figure 2.3: Effect of data blocking during a VMC simulation. Note that the block averages fluctuate much less than the individual estimates of E_L.

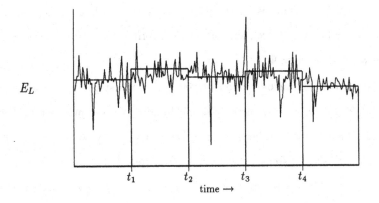

equal σ^2. This procedure can be implemented by using an ensemble of independently selected walkers. Once n such walkers have been equilibrated, Eq. 2.29 can again be used, where m is the number of steps in the walk and $\langle A \rangle_i$ is the average value of the i'th independent walker over these steps.

2.3 Other Sampling Methods

In previous sections, a property such as $\langle A \rangle_{\Psi^2}$ was sampled by summing A over points drawn from the distribution Ψ^2. In this section we outline sampling methods designed to improve the estimation of A in specific cases. In particular, we introduce the use of an auxiliary guiding function and methods of correlated sampling of differences.

2.3.1 Guiding functions

The efficiency of a Monte Carlo evaluation of a quantity such as the local energy, E_L, can be characterized by the variance achieved with a given amount of computation time. The efficiency depends on Ψ because the Monte Carlo error depends on the variance of the function $E_L(\mathbf{x})$, and on the serial correlation between steps in a walk, both of which depend on Ψ.

Serial correlation can be reduced by increasing the step size — but, as noted before, we can only increase the step so much before the acceptance ratio becomes too small. We can, however, increase the *effective* step size by evaluating the function of interest only after m steps rather than at every step. (This procedure is similar to data blocking.) The sample mean is unchanged; however, the variance has changed since we have reduced the correlation between measurements. We also eliminate a large amount of redundant data.

The Monte Carlo method is most effective when combined with the use of a *guiding function*, Ψ_G. A guiding function is a function chosen to imitate Ψ, but that is less costly to evaluate (in computation time). The guiding function is used to generate a distribution Ψ_G^2. By then reweighting one can obtain averages with respect to Ψ^2, i.e.,

$$\langle E_L \rangle_{\Psi^2} = \frac{\int \Psi_G^2(\mathbf{x})\, w(\mathbf{x}) E_L(\mathbf{x})\, d\mathbf{x}}{\int \Psi_G^2(\mathbf{x}) w(\mathbf{x})\, d\mathbf{x}} = \lim_{M \to \infty} \frac{\sum_{i=1}^M E_L(\mathbf{X}_i) w(\mathbf{X}_i)}{\sum_{i=1}^M w(\mathbf{X}_i)} \ . \tag{2.30}$$

The quantity $w(\mathbf{X}_i) = \Psi^2(\mathbf{X}_i)/\Psi_G^2(\mathbf{X}_i)$ is a weight which converts the distribution from Ψ_G^2 to Ψ^2. Even though Eq. 2.30 reduces formally to the average over Ψ^2, the roles of Ψ_G and Ψ are different. The inexpensive function Ψ_G is used at every step of the walk, whereas the relatively expensive, but more accurate, Ψ is used to obtain the required averages and is computed only every m steps.

To see the benefit of a guiding function, suppose that the correlation between points one step apart, ρ_1, is near unity, while ρ_m, the correlation after $m > K_0$ steps, is effectively zero. From Eq. 2.28, after N steps, with E_L evaluated at every step, the variance of the mean of E_L is

$$\sigma^2(1) = \frac{\hat{\sigma}^2}{N} \left(1 + 2 \sum_{k=1}^{m} \rho_k \right) \qquad (2.31)$$

where $\hat{\sigma}^2$ is here the variance of the distribution $E_L(\mathbf{x})$. Alternatively, by evaluating E_L only every m steps, we reduce the number of samples from N to N/m, but we also decrease the serial correlation term to zero, so that

$$\sigma^2(m) = \frac{m\hat{\sigma}^2}{N}. \qquad (2.32)$$

Comparing Eqs. 2.31 and 2.32, we see that their ratio is

$$\frac{\sigma^2(m)}{\sigma^2(1)} = \frac{m}{(1 + 2\sum_{k=1}^{m} \rho_k)}. \qquad (2.33)$$

The sum over all ρ_k between 1 and m must be less than m. Hence, the smallest value of the ratio $\sigma^2(m)/\sigma^2(1)$ is about $1/2$, which occurs when $\rho_k = 1$ for all k up to m. The largest value of this ratio, namely $m/3$, results if $\rho_1 = 1$ and $\rho_k = 0$ ($k \neq 1$). Neither is realistic though, and we are better off with an intermediate guess of $\rho_k \approx 1/2$ for all k up to m, leading to $\sigma^2(m)/\sigma^2(1) \cong 1$.

So far there appears to be no real gain. However, if Ψ_G can be computed with g times less cost than Ψ, E_L can be determined g-times more often than indicated in Eq. 2.32. If we assume that nearly all the computation time is taken up in the evaluation of Ψ, and very little is consumed by moving the walkers, then we may carry this factor through to Eq. 2.33 to obtain,

$$\frac{\sigma^2(m)}{\sigma^2(1)} = \frac{m}{g(1 + 2\sum_{k=1}^{m} \rho_k)} \approx 1/g, \qquad (2.34)$$

for equal computation times. Although this is only a rough estimate, it is sufficient to show that use of a guiding function can be very beneficial. In a real application, we must realize that the overhead due to moving walkers may be comparable to that to compute Ψ_G, and thus we do not expect a full factor of g. In addition, the best value of m may not be K_0. Rather, it should be determined empirically.

One has much flexibility in choosing Ψ_G. Care must be taken, however, to insure that the weights $w = \Psi^2/\Psi_G^2$ are well behaved. In particular, either Ψ_G^2 must be nonzero everywhere, or at most vanish no faster than Ψ^2 wherever Ψ^2 vanishes. One should also consider the trade off between complexity and accuracy of both Ψ and Ψ_G. Since the variance of a Monte Carlo run depends not only on the inherent variance of the measured property, but also on the number of samples one takes, the "complexity" of the wave function affects the variance in both a good and a bad way. Let us analyze this trade-off.

Let T be a set amount of computation time. Consider two functions Ψ_1 and Ψ_2 requiring times T_1 and T_2, respectively, to move the ensemble one step. The respective variances of E_L are σ_1^2 and σ_2^2. The Monte Carlo errors are the standard deviation of the means, and since we can sample from Ψ_1, $N_1 = T/T_1$ times, and similarly for Ψ_2 these errors are $\epsilon_1 = \sqrt{T_1\sigma_1^2/T}$ and $\epsilon_2 = \sqrt{T_2\sigma_2^2/T}$. The ratio of errors is

$$\frac{\epsilon_1}{\epsilon_2} = \frac{\sigma_1}{\sigma_2}\sqrt{\frac{T_1}{T_2}}. \tag{2.35}$$

Here the trade-off between a smaller σ and a larger T is seen explicitly. To analyze Eq. 2.35 further, we can make some assumptions about the factors contributing to T. Let $T_1 = a_1 + b_1 s_1$ and $T_2 = a_2 + b_2 s_2$, where s_1 and s_2 represent the number of operations needed to evaluate Ψ_1 and Ψ_2, respectively, and a_1, b_1, a_2, and b_2 are constants associated with the algorithm used (i.e., a_1 and a_2 are related to the overhead of moving the walkers, and b_1 and b_2 are related to the expense of wave function evaluation). Assume that Ψ_1 and Ψ_2 are similar in such a way that $a_1 = a_2 = a$ and $b_1 = b_2 = b$ so that only the *number* of operations differs. Define $c \equiv b/a$. This gives $T_1/T_2 = (1 + cs_1)/(1 + cs_2)$. There are two limits we can look at. In the limit where cs_1 and cs_2 are much smaller than unity, i.e., overhead dominates the cost, then $T_1/T_2 \cong 1$. The implication of this result is that if an algorithm is dominated by steps *other* than the evaluation of Ψ, then one should always use a better Ψ and thereby reduce σ. Alternatively, if cs_1 and cs_2 are large, then $T_1/T_2 \cong s_1/s_2$, and the algorithm is dominated by the calculation of Ψ. In this case it is less clear what to choose. We must evaluate the σ_1/σ_2 contribution in Eq. 2.35 and compare it to $\sqrt{T_1/T_2}$. Note, however, that if greater accuracy is the main concern, then more complexity may be required regardless of the computational cost.

2.3.2 Correlated sampling

In Sec. 2.2.5 we discussed methods to reduce the effects of serial correlation. In this section we will use serial correlation to our advantage to evaluate energy differences. Having introduced the guiding function, Ψ_G, we can go one step further and introduce new trial functions, $\Psi_0, \Psi_1, \cdots \Psi_n$, each slightly different from the others. Such a set of functions can be generated whenever a central function, say Ψ_0, is perturbed in a number of different ways. Examples of such perturbations are electric fields, molecular geometry displacements, and changes in the variational parameters defining Ψ_0. Then from energy calculations with this set of $\{\Psi_i\}$, one can determine dipole moments and other response functions, optimize geometries, or optimize Ψ_0 itself (e.g., with respect to the energy). In addition, although we focus here on energy differences, the same method can be applied to differences in other properties of interest.

Define $E_L^{(i)}$ to be the local energy of wave function Ψ_i. The expectation value of all the $E_L^{(i)}$ can then be determined simultaneously by substituting $E_L^{(i)}$ and $w^{(i)} = \Psi_i^2/\Psi_G^2$ in the appropriate places in Eq. 2.30. Because the $E_L^{(i)}$ are computed over the same random positions, $\mathbf{X}_k$, due to the correlation between them, the difference

$$\langle E_L^{(i)} \rangle - \langle E_L^{(j)} \rangle = \frac{\int \Psi_G^2 w^{(i)} E_L^{(i)} \mathrm{d}\mathbf{x}}{\int \Psi_G^2 w^{(i)} \mathrm{d}\mathbf{x}} - \frac{\int \Psi_G^2 w^{(j)} E_L^{(j)} \mathrm{d}\mathbf{x}}{\int \Psi_G^2 w^{(j)} \mathrm{d}\mathbf{x}} \qquad (2.36)$$

will have lower variance than if the the two energies were computed from different random walks. In the literature this procedure is known as correlated sampling or differential Monte Carlo.

Variance reduction by this method can be quite large. Consider the problem of computing the derivative of the energy with respect to an electric field perturbation, F_x, in order to obtain the dipole moment. Let the Monte Carlo computed energy of the unperturbed system be $E(0)$ and the energy of the perturbed system be $E(F_x)$, with associated variances designated $\sigma^2(0)$ and $\sigma^2(F_x)$, respectively. The finite difference approximation to dE/dF_x is $\Delta E/F_x$, where $\Delta E \equiv E(F_x) - E(0)$. However, the variance of the derivative is $\sigma_d^2 \equiv \sigma^2(\Delta E)/F_x$, where $\sigma^2(\Delta E)$ is the variance of the quantity ΔE over the Monte Carlo sample. As $F_x \to 0$, the finite difference approximation approaches the exact derivative. One concern is that σ_d^2 might diverge in this limit. From Ch. 1 we recall that the variance is given by

$$\sigma^2(\Delta E) = \sigma^2(F_x) + \sigma^2(0) - 2\,\mathrm{Cov}(E(F_x), E(0)). \qquad (2.37)$$

In the limit $F_x \to 0$, $E(F_x) \to E(0)$ and both $\sigma(F_x)$ and $\mathrm{Cov}(E(F_x), E(0)) \to \sigma(0)$. Hence with correlated sampling $\sigma^2(\Delta E)$ also goes to zero, and the variance of the derivative remains finite.

In Table 2.3 we show energy gradients with respect to nuclear coordinates computed using correlated sampling.

2.4 Monte Carlo Optimization

Optimization lies at the root of all *ab initio* methods — from geometry optimization of a molecule to parameter optimization of an analytic wave function. In this section

Table 2.3: VMC energies, first and second derivatives of H_2 and Li_2 obtained using correlated sampling.[10]

System	r(bohr)	Value	E	dE/dr	d^2E/dr^2
H_2	0.9	VMC	−1.08185(4)	−0.5094(35)	2.42(5)
	0.9	Exact	−1.0836	−0.5007	2.26
	1.4011	VMC	−1.17217(4)	0.0048(10)	0.43(3)
	1.4011	Exact	−1.17447	0.0000	0.37
	1.9	VMC	−1.14186(5)	0.0860(9)	0.098(7)
	1.9	Exact	−1.14685	0.0852	0.091
Li_2	3.5	VMC	−14.9165(5)	−0.062(4)	0.082(7)
	3.5	Exact	−14.9598	−0.064	0.086
	5.05	VMC	−14.9564(6)	−0.0008(4)	0.024(7)
	5.05	Exact	−14.9967	0.00017	0.017

we will be concerned primarily with optimization of Ψ to achieve the greatest accuracy. All the methods presented, however, can also be used for other optimizations, including locating equilibrium geometries.

In variational basis set methods, the wave function depends on basis function parameters, and the energy expression is minimized to determine these parameters. Even in perturbation theory, the zeroth-order wave function is typically obtained in this manner. This need to minimize the energy with respect to a set of parameters is a common feature of *ab initio* methods. Typically the wave function depends on linear combinations of basis functions. The resulting linear equations are then amenable to matrix algebra methods. (However, basis set exponents are non-linear; see Ch. 5.) Because of the freedom provided by Monte Carlo, one has numerous options for the wave function form, which may involve many linear *and* non-linear parameters. The challenge lies in the optimization of the many nonlinear parameters using Monte Carlo computed quantities.

Figure 2.4: Dependence of the variational energy and variance of the local energy on α for the simple H_2 trial function of Eq. 2.5.

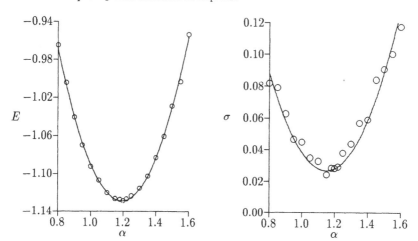

2.4.1 Optimization criteria

Although Monte Carlo methods can be used to optimize the energy with respect to a set of variational parameters, there is a certain advantage to optimizing the variance of the local energy. As shown earlier, for the exact wave function

$$\mathrm{var}(E[\Phi]) = 0. \tag{2.38}$$

One advantage of minimizing the variance of the local energy is that the minimum of the variance (namely zero) is known *a priori*, whereas the minimum energy is not known. Figure 2.4 shows the dependence of $E[\Psi]$ and σ on α for the simple Ψ of Eq. 2.5. For this example we see that the two curves both have their minima at $\alpha \sim 1.2$.

2.4.2 Energy optimization

In Monte Carlo optimization, a principle concern is the statistical uncertainty of the quantity being optimized. The magnitude of this uncertainty will have great impact on the convergence as well as on the ability to actually find the optimum parameter set. Consider the following simple energy minimization scheme: choose a set of parameters $\{\alpha\}$; compute the energy; vary $\{\alpha\}$; compare the new energy to the old; and retain the set that yields the lowest energy. Repeat this procedure until the energy no longer changes to within a preassigned value. The problem is that if E is obtained by Monte Carlo, the energy E_1 associated with parameters $\{\alpha\}_1$ has a statistical error of ϵ_1, while E_2 computed with $\{\alpha\}_2$ has a statistical error ϵ_2. In other words, one cannot be certain which energy is lower. If one uses 95% confidence intervals, one only knows that the exact energies $\tilde{E}_1$ and $\tilde{E}_2$ of $\{\alpha\}_1$ and $\{\alpha\}_2$ are in the ranges

$$E_1 - 2\epsilon_1 \;\; \leq \;\; \tilde{E}_1 \;\; \leq \;\; E_1 + 2\epsilon_1 \tag{2.39}$$
$$E_2 - 2\epsilon_2 \;\; \leq \;\; \tilde{E}_2 \;\; \leq \;\; E_2 + 2\epsilon_2.$$

Deciding which set of parameters is optimal can only be answered statistically.

The quantity of interest is neither $\tilde{E}_1$ nor $\tilde{E}_2$, but the difference $\tilde{E}_1 - \tilde{E}_2$. Therefore, we might consider using correlated sampling to address the problem. To do so, let us first choose $\Psi\{\alpha\}_0$ as a guiding function. The energies associated with many sets of closely related parameters $\{\alpha\}_1 \ldots \{\alpha\}_m$ can then be evaluated simultaneously. The variance of the energy differences will be lower than that obtained by combining the confidence intervals as long as the energies are correlated. Hence this provides an advantageous method of optimizing $E[\Psi\{\alpha\}]$. With this method one can also compute second differences for use in a Newton-Raphson or other second derivative method.

Alternatively, the gradient of the energy with respect to $\{\alpha\}$, i.e. $\nabla_\alpha E$, can be computed *analytically* at each step in the walk. Is there any advantage of analytically-computed over finite-difference-computed derivatives? If it were not for the correlated sampling method, the answer would be that analytic gradients are always preferred. The deciding factors here must be the relative variance of the analytically computed gradient versus that of the energy differences, and conditions of numerical accuracy versus simplicity of programming and speed. If the analytic derivatives contain too much statistical uncertainty, or take too much time to compute, then correlated sampling may be the better choice.

Regardless of which method is used, for *a given sample* $\{\mathbf{X}_k\}$, the Monte Carlo energy can be either above or below $E[\Psi\{\alpha\}]$. Energy minimization therefore requires very accurate energy difference estimates. Otherwise, it is possible that the optimal values of $\{\alpha\}$ may vary so greatly from sample to sample that convergence cannot be obtained.

2.4.3   Variance optimization

The *a priori* lower bound (of zero) for the variance is a major advantage. Formally, the variance of the local energy is

$$\sigma^2(E_L) = \int (E_L(\mathbf{x}) - E[\Psi\{\alpha\}])^2 \, \Psi^2 d\mathbf{x}. \tag{2.40}$$

The Monte Carlo estimator for this variance over a given sample $\{\mathbf{X}\}$ is denoted by $\sigma^2(E_L\{\mathbf{X}\})$ as an explicit reminder of the sample dependence. One estimate of $\sigma^2(E_L\{\mathbf{X}\})$ is the variance over the finite sample, namely, $\langle E_L^2 \rangle - \langle E_L \rangle^2$. This quantity will be a good approximation to σ^2 only if the average over the finite sample $\langle E_L\{\mathbf{X}\} \rangle$ is close to $E[\Psi\{\alpha\}]$. To avoid this problem, many workers have chosen instead to optimize,

$$\sigma_d^2(E_L\{\mathbf{X}\}) = \frac{1}{N}\sum_{i=1}^{N}(E_L(\mathbf{X}_i) - E_{\text{ref}})^2. \qquad (2.41)$$

Here E_{ref} is a parameter set by the practitioner to be as close to the best optimized value of $E[\Psi\{\alpha\}]$ as possible. As the optimization converges, E_{ref} can be updated.

Equation 2.41 is in the form of a discrete, least-squares fitting of data $(E_L\{\mathbf{X}\})$ to a function E_{ref}. Note that if E_{ref} is not equal to $E[\Psi\{\alpha\}]$, which is not known exactly, then Eq. 2.41 mixes both variance and energy estimators. This can be seen by setting $E_{\text{ref}} \equiv E[\Psi\{\alpha\}] + \delta E$. Then Eq. 2.41 becomes

$$
\begin{aligned}
\sigma_d^2(E_L\{\mathbf{X}\}) &= \frac{1}{N}\sum_{i=1}^{N}(E_L(\mathbf{X}_i) - E[\Psi\{\alpha\}] - \delta E)^2 \\
&= \frac{1}{N}\sum_{i=1}^{N}\{(E_L(\mathbf{X}_i) - E[\Psi\{\alpha\}])^2 - 2\delta E(E_L(\mathbf{X}_i) - E[\Psi\{\alpha\}]) + \delta E^2)\} \\
&= \sigma^2(E_L(\mathbf{X}_i)) + \delta E^2 - 2\delta E(\langle E_L\rangle - E[\Psi\{\alpha\}]). \qquad (2.42)
\end{aligned}
$$

In the final expression only the first term, $\sigma^2(E_L(\mathbf{X}_i))$, is the quantity we intended to minimize. The second term is a positive constant which does not affect the minimization. The final term, however, depends linearly on the local energy. This term can be made small by choosing E_{ref} close to $E[\Psi\{\alpha\}]$, (i.e., δE small), or by using a large enough sample so that $\langle E_L\rangle = E[\Psi\{\alpha\}]$. Alternatively, we can use this final term to drive the optimization to find both lower variance *and* lower energy, by setting E_{ref} somewhat lower than $E[\Psi\{\alpha\}]$, making δE negative.

We may now optimize the estimate of the variance in a similar manner to the energy optimization described earlier. Generate a sample $\{\mathbf{X}\}$ from which to determine σ_d^2 and, if possible, $\nabla_\alpha\sigma_d^2$. The gradient can be obtained by either analytical or correlated sampling methods. Then a suitable optimization scheme can be used (e.g., steepest descent, Newton's method, etc.). Note that any optimization based upon the sample $\{\mathbf{X}\}$ will be biased toward that sample unless the sample size is very large. Thus one should always test the optimization's convergence by generating several independent samples. Because of this sample dependence, one may decide not to optimize $\{\alpha\}$ entirely with respect to $\{\mathbf{X}\}$, but rather to choose $\{\alpha\}_1$ based upon sample $\{\mathbf{X}\}_1$, then generate a new sample $\{\mathbf{X}\}_2$ from $\Psi^2\{\alpha\}_1$ to determine $\{\alpha\}_2$, and so forth. The implementation used to optimize $\Psi(\{\alpha\})$, i.e., size of sample and optimization algorithm, may therefore depend upon the magnitude of this sample dependence which is observed in the system in question. For cases with little sample dependence, any optimization scheme may work. For cases with large sample dependence, special care must be taken to insure that a true minimum has been obtained. Variance optimization methods have been used extensively to generate relatively complicated forms of Ψ. In Table 2.4 we give several representative examples. We see that together with optimization, the VMC method can produce very impressive results.

Exercises

1. Prove that the variational energy is an upper bound to the exact energy by expanding Ψ in the eigenstates of $\mathcal{H}$, and examining the resulting energy expression. Assume Ψ is normalized.

2. Write a computer program to carry out the simple Metropolis walk of Sec. 2.2.1 for a particle in a box. How would you handle the boundary conditions?

3. Write a computer program for H_2, using the simple Ψ of Eq. 2.5 following Algorithm 2.1. Find the optimum value of α for an H-H distance of 1.4 bohr.

4. Show that the operator $G = e^{-\mathcal{L}\delta t}$ (i.e., Eq. 2.21) is a solution of Eq. 2.20. Assuming that the force $\mathbf{F}$ remains constant from $\mathbf{x}$ to $\mathbf{y}$, integrate over a small time interval to obtain Eq. 2.22 from Eq. 2.21.

Table 2.4: Total energies $\langle E_L \rangle$ and standard deviations σ of optimized wave functions by VMC.[11]

System	$\langle E_L \rangle$	$\sigma(E_L)$	Exact
He	$-2.903726(4)$	0.0011	-2.903724
Li$^-$	$-7.4998(1)$	0.045	-7.5004
Be	$-14.6664(3)$	0.092	-14.6673
Ne	$-128.884(4)$	1.0	-128.937

5. Using the program of either Exercise 2 or 3, compute and plot ρ_k for various time steps as was done in Fig. 2.2.

Suggestions for Further Reading

1. A. C. Hurley, *Introduction to the Electronic Theory of Small Molecules* (Academic, New York, 1976).

2. D. M. Ceperley, G. V. Chester and M. H. Kalos, "Monte Carlo Simulation of a Many-Fermion System," *Physical Review B* **16**, 3081-99 (1977).

3. D. M. Ceperley and M. H. Kalos, "Quantum Many-Body Problems," in *Monte Carlo Methods in Statistical Physics*, 145-94 (Springer-Verlag, New York, 1979).

4. J. W. Moskowitz and M. H. Kalos, "A New Look at Correlations in Atomic and Molecular Systems. I. Applications of Fermion Monte Carlo Variational Method," *International Journal of Quantum Chemistry* **20**, 1107-119 (1981).

5. T. Yoshida and K. Iguchi, "Variational Monte Carlo Method in the Connected Moments Expansion: H, H$^-$, Be, and Li$_2$," *Journal of Chemical Physics* **91**, 4249-53 (1989).

6. H. L. Gordon, S. M. Rothstein and T. R. Proctor, "Efficient Variance-Reduction Transformations for the Simulation of a Ratio of Two Means: Application to Quantum Monte Carlo Simulations," *Journal of Computational Physics* **47**, 375-86 (1982).

References

1. E. A. Hylleraas, "Neue Berechnung der Energie des Heliums in Grand-austande, sowie des tiefsten terms von ortho-Helium," *Zeishrift für Physik* **54**, 347-66 (1929).

2. C. L. Pekeris, "Ground State of Two-Electron Atoms," *Physical Review* **112**, 1649-58 (1958).

3. J. Lee, "The Upper and Lower Bounds of the Ground State Energies Using the Variation Method," *American Journal of Physics* **55**, 1039-40 (1987).

4. K. E. Schmidt and J. W. Moskowitz, "Correlated Monte Carlo Wave Functions for the Atoms He through Ne," *Journal of Chemical Physics* **93**, 4172-78 (1990).

5. E. Clementi and C. Roetti, *Atomic Data and Nuclear Data Tables* **14**, 177-478 (1974).

6. A. Veillard and E. Clementi, "Correlation Energy in Atomic Systems. V. Degeneracy Effects for Second Row Atoms," *Journal of Chemical Physics* **49**, 2415-21 (1962).

7. P. J. Reynolds, R. N. Barnett, B. L. Hammond and W. A. Lester, Jr., "Molecular Physics and Chemistry Applications of Quantum Monte Carlo," *Journal of Statistical Physics* **43**, 1017-26 (1986).

8. A. D. McLean and M. Yoshimine, "Molecular Properties Which Depend on the Square of Electronic Coordinates: H_2 and HNO," *Journal of Chemical Physics* **45**, 3676-81 (1966).

9. W. Kolos and L. Wolniewicz, "Potential Energy Curves for the $X\ ^1\Sigma_g^+$, $b\ ^3\Sigma_u^+$ and $C\ ^1\Pi_u$ States of the Hydrogen Molecule," *Journal of Chemical Physics* **43**, 2429-41 (1965).

10. Z. Sun, W. A. Lester, Jr. and B. L. Hammond, "Correlated Sampling of Monte Carlo Derivatives with Iterative-fixed Sampling," *Journal of Chemical Physics* **97**, 7585-89 (1992).

11. C. J. Umrigar, K. G. Wilson and J. W. Wilkins, "A Method for Determining Many Body Wavefunctions," in *Computer Simulation Studies in Condensed Matter Physics: Recent Developments*, Springer Proceedings in Physics (Springer, Berlin, 1988).

Chapter 3

Green's Function Methods

In Chapter 2 we showed how to use Monte Carlo methods to evaluate quantum mechanical expectation values associated with a trial wave function Ψ. The results depended entirely upon the choice of the form of Ψ. If highly accurate results are sought, the question of how to specify Ψ to assure a desired accuracy is not straightforward. Post-Hartree-Fock methods that use 10^5 to 10^6 determinants can yield 99% of the correlation energy for small systems involving first row atoms. To obtain 99.9% of the correlation energy for these same systems requires several orders of magnitude more determinants, as well as either a very large one-electron basis set or, for atoms and diatomics, numerical molecular orbitals. More general forms of Ψ's involving interelectronic coordinates converge much faster — as was shown by Hylleraas and others (cf. Ch. 5) for He. The problem is that for many-electron polyatomic systems, such wave functions lead to integrals that present insurmountable difficulties for conventional methods.

An important alternative is to adopt a numerical or simulational method that eliminates the need for constructing and optimizing the wave function. Such methods include grid methods and quantum Monte Carlo methods (QMC). For solving the Schrödinger equation Monte Carlo methods have a clear advantage over grid methods because of the high dimensionality of the position space of the electrons.

To solve a differential equation with Monte Carlo methods, one first constructs the corresponding integral equation. This transformation involves the determination of the Green's function of the system in question. The focus of this chapter will be formulating and implementing Green's functions that will provide us with a solution to the time-independent Schrödinger equation.

At first sight QMC Green's function methods may seem very different from the previously discussed VMC method. As we will see however, the two are closely connected. We shall show that QMC encompasses VMC, diffusion Monte Carlo (DMC), and methods where the exact Green's function is sampled (often called GFMC).

Use of Green's functions in quantum mechanics is of course not restricted to the current context. Green's functions are encountered in a number of methods (for example, in electron propagator methods). The unique feature here is the use of Monte Carlo methods both to sample the configuration space of the electrons and in some cases also to sample the Green's function itself. We begin with a general discussion of Green's functions as applied to the solution of the Schrödinger equation. Following that we will derive approximate Green's functions and discuss the DMC algorithm with importance sampling. Exact Green's function methods are developed in Sec. 3.3.

3.1   Integral Form of Schrödinger's Equation

We now discuss the problem of solving the Schrödinger equation directly using the Monte Carlo method. Specifically, we will focus for now on the ground eigenstate Φ_0 with eigenvalue E_0 which satisfies the time-independent Schrödinger equation,

$$\mathcal{H}|\Phi_0\rangle = E_0|\Phi_0\rangle, \tag{3.1}$$

where $\mathcal{H}$ is the molecular Hamiltonian, composed of the kinetic energy operator $\mathcal{T}$ (proportional to the Laplacian operator) and the multiplicative potential operator $\mathcal{V}$. The Monte Carlo method is not well suited for solving differential equations directly. Rather, as shown in the previous chapters, the method is useful for

- creating iteratively a Markov chain of states;

- estimating integrals.

In connection with these capabilities, let us recast Eq. 3.1 into an iterative integral equation.

3.1.1 Green's function formalism

The formal solution of Eq. 3.1 can be obtained by operating on Eq. 3.1 from the left by the inverse operator $\mathcal{H}^{-1}$:

$$|\Phi_0\rangle = E_0 \mathcal{H}^{-1} |\Phi_0\rangle. \tag{3.2}$$

Because $\mathcal{H}$ is a differential operator, $\mathcal{H}^{-1}$ is an integral operator. To deduce the form of the latter, introduce a complete set of position states, $\int_{-\infty}^{\infty} |\mathbf{x}\rangle\langle\mathbf{x}| d\mathbf{x}$, between $\mathcal{H}^{-1}$ and Φ_0, and multiply from the left by the position state $\langle\mathbf{y}|$ to obtain

$$\langle\mathbf{y}|\Phi_0\rangle = E_0 \int_{-\infty}^{\infty} \langle\mathbf{y}|\mathcal{H}^{-1}|\mathbf{x}\rangle\langle\mathbf{x}|\Phi_0\rangle \, d\mathbf{x}. \tag{3.3}$$

Define the *Green's function* associated with $\mathcal{H}$ to be

$$G(\mathbf{y}, \mathbf{x}) \equiv \langle\mathbf{y}|\mathcal{H}^{-1}|\mathbf{x}\rangle, \tag{3.4}$$

which is the position-space representation of $\mathcal{H}^{-1}$. Then Eq. 3.3 becomes the integral equation,

$$\Phi_0(\mathbf{y}) = E_0 \int_{-\infty}^{\infty} G(\mathbf{y}, \mathbf{x})\Phi_0(\mathbf{x}) d\mathbf{x}. \tag{3.5}$$

We may now attempt to solve Eq. 3.5 iteratively by choosing some initial state $\phi^{(0)}$ and a trial energy E_T which approximate Φ_0 and E_0 respectively, and then form the series,

$$\phi^{(n+1)}(\mathbf{y}) = E_T \int\limits_{-\infty}^{\infty} G(\mathbf{y}, \mathbf{x}) \phi^{(n)}(\mathbf{x}) \, d\mathbf{x}. \tag{3.6}$$

The convergence of this series will be examined below, but first we consider how to specify $G(\mathbf{y}, \mathbf{x})$.

By definition,

$$\mathcal{H}(\mathcal{H}^{-1})\Phi_0 = \Phi_0. \tag{3.7}$$

Using Eqs. 3.2 and 3.5 then leads to

$$\Phi_0(\mathbf{y}) = \mathcal{H}(\mathbf{y})\mathcal{H}^{-1}\Phi_0(\mathbf{y}) = \int\limits_{-\infty}^{\infty} \mathcal{H}(\mathbf{y}) G(\mathbf{y}, \mathbf{x})\Phi_0(\mathbf{x}) \, d\mathbf{x} \ ,$$

where $\mathcal{H}(\mathbf{y})$ indicates that the Laplacian and potential energy operators are to be taken at point $\mathbf{y}$. Equation 3.8 thus implies

$$\mathcal{H}(\mathbf{y}) G(\mathbf{y}, \mathbf{x}) = \delta(\mathbf{y} - \mathbf{x}). \tag{3.8}$$

The reverse also must hold, i.e.

$$\mathcal{H}(\mathbf{x}) G(\mathbf{y}, \mathbf{x}) = \delta(\mathbf{y} - \mathbf{x}). \tag{3.9}$$

The substitution of either Eq. 3.8 or Eq. 3.9 into Eq. 3.5 returns the original form of the Schrödinger equation, Eq. 3.1.

The Green's function of a Hermitian operator, such as the Hamiltonian, is symmetric, i.e. $G(\mathbf{y}, \mathbf{x}) = G(\mathbf{x}, \mathbf{y})$. One can show this property by retracing the steps leading to Eq. 3.4 starting from the complex conjugate of Eq. 3.2,

$$\langle\Phi_0| = E_0\langle\Phi_0|\mathcal{H}^{-1} \tag{3.10}$$

rather than Eq. 3.2 (which is possible because Hermitian operators can act to the left or right). Inserting a complete set of position states between $\mathcal{H}^{-1}$ and Φ_0, as was done to obtain Eq. 3.3, yields

$$\Phi_0^*(\mathbf{y}) = E_0 \int\limits_{-\infty}^{\infty} G(\mathbf{x},\mathbf{y})\Phi_0^*(\mathbf{x})\,d\mathbf{x}. \tag{3.11}$$

Comparing Eq. 3.5 with 3.11 we see that $G(\mathbf{y},\mathbf{x})$ must equal $G(\mathbf{x},\mathbf{y})$ for real states $\Phi_0 = \Phi_0^*$.

In the Monte Carlo applications we discuss here, $G(\mathbf{y},\mathbf{x})$ must be positive everywhere and normalizable, so that it may be interpreted as a transition probability. Then Eq. 3.6 may be generated by a random walk process in which $G(\mathbf{y},\mathbf{x})$ is the probability of moving from $\mathbf{x}$ to $\mathbf{y}$. This gives Eq. 3.5 the following interpretation: for each point $\mathbf{y}$ the wave function is composed of the sum over all other points $\mathbf{x}$ weighted by the probability of a particle moving from $\mathbf{x}$ to $\mathbf{y}$ (and normalized by E_0). Unlike the original differential form of the Schrödinger equation, in which the wave function at a point is determined only by information *local* to that point (i.e., the curvature and potential), Eq. 3.5 relates the wave function at a point, *globally* to all other points.

Consider the example of a particle in a one-dimensional box with walls at 0 and L. The Green's function in the box is given by

$$\frac{d^2}{dx^2}G(y,x) = \delta(y - x). \tag{3.12}$$

There is an equivalent relation for the derivative with respect to y. Outside the box the Green's function is zero. For $x \neq y$, Eq. 3.12 is satisfied by any linear function of x and y, but how can the delta function be produced? What is needed is a discontinuous first derivative at $x = y$. The Green's function which satisfies these conditions is

$$G(y,x) = \begin{cases} \frac{1}{2}(L-x)y & \text{if } y \leq x \\ \frac{1}{2}(L-y)x & \text{if } y \geq x \\ 0 & \text{if } x,y < 0 \text{ or } > L. \end{cases} \tag{3.13}$$

Figure 3.1: Green's function for the one-dimensional particle in a box.

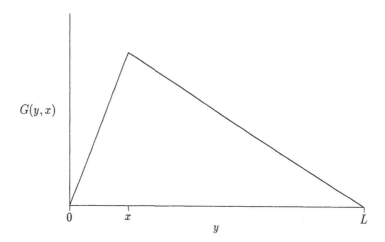

This Green's function is shown in Fig. 3.1 and is the subject of Exercise 3.1.

Looking at Eq. 3.8 one might think that we have achieved nothing, because it is just as difficult to solve as was Eq. 3.1. However, we will show in later sections that the Green's function either may be approximated or it may be sampled by Monte Carlo methods.

To understand the action of the Green's function, let us expand it in eigenstates of $\mathcal{H}$,

$$G(\mathbf{y}, \mathbf{x}) = \sum_{k=0}^{\infty} b_k \Phi_k(\mathbf{x}) \Phi_k(\mathbf{y}), \qquad (3.14)$$

and solve for b_k. Substituting Eq. 3.14 into the left-hand side of Eq. 3.8 yields

$$\mathcal{H}G(\mathbf{y}, \mathbf{x}) = \sum_{k=0}^{\infty} b_k E_k \Phi_k(\mathbf{x}) \Phi_k(\mathbf{y}). \qquad (3.15)$$

Similarly, the delta function on the right-hand side of Eq. 3.8 may be expanded as

$$\delta(\mathbf{x} - \mathbf{y}) = \sum_{k=0}^{\infty} \Phi_k(\mathbf{x}) \Phi_k(\mathbf{y}). \qquad (3.16)$$

By comparing terms between Eqs. 3.15 and 3.16 we conclude that $b_k = E_k^{-1}$, and therefore

$$G(\mathbf{y}, \mathbf{x}) = \sum_{k=0}^{\infty} E_k^{-1} \Phi_k(\mathbf{x}) \Phi_k(\mathbf{y}).$$ (3.17)

Substitution of this into Eq. 3.5 shows the validity of Eq. 3.17.

The convergence properties of Eq. 3.6 can now be seen by expanding $\phi^{(0)}$ in eigenstates of $\mathcal{H}$, and using Eq. 3.17 for G. The result is

$$\phi^{(n)} = \sum_{k=0}^{\infty} \left[\frac{E_T}{E_k}\right]^n c_k \Phi_k,$$ (3.18)

where c_k are the expansion coefficients of $\phi^{(0)}$. If $|E_0| < |E_1| < |E_2|$, etc., then $\phi^{(\infty)}$ will be dominated by the ground state. For electronic structure, we are mainly interested in the negative total energy states for which $|E_0| > |E_1| > |E_2|$. In this case we can invert the eigenvalue spectrum by subtracting a constant energy offset, $E_{\min} < E_0$, from both sides of Eq. 3.1. This subtraction results in replacing each energy in this discussion with $E_k - E_{\min}$.

The n-th iteration of Eq. 3.6 would then yield

$$\phi^{(n)} = \sum_{k=0}^{\infty} \left[\frac{E_T - E_{\min}}{E_k - E_{\min}}\right]^n c_k \Phi_k.$$ (3.19)

Because $E_T - E_{\min} > 0$, we have

$$\frac{E_T - E_{\min}}{E_0 - E_{\min}} > \frac{E_T - E_{\min}}{E_1 - E_{\min}} > \frac{E_T - E_{\min}}{E_2 - E_{\min}} > \ldots.$$ (3.20)

Thus in the limit $n \to \infty$, $\phi^{(n)}$ converges to the lowest state with a nonzero coefficient, that is, the lowest state not orthogonal to $\phi^{(0)}$. Application of this method to electronic structure is presented in Sec. 3.3.1 and Sec. 3.3.2.

3.1.2 Time-dependent Green's functions

Let us now turn to an alternative procedure for finding solutions to the time-independent Schrödinger equation by considering the *time-dependent* Schrödinger equation, i.e.,

$$-\frac{\partial\phi(\mathbf{x},t)}{i\partial t} = (\mathcal{H} - E_T)\phi(\mathbf{x},t). \tag{3.21}$$

(E_T is an arbitrary energy shift and $\hbar = 1$ in atomic units. See Appendix A.) Our motivation is to exploit the formal similarity between Eq. 3.21 (in imaginary time) and the classical diffusion equation. This will be discussed at greater length in Sec. 3.2. The formal solution of Eq. 3.21 may be expressed in terms of the eigenfunctions and eigenvalues of $\mathcal{H}$,

$$\phi(\mathbf{x},t) = \sum_{k=0}^{\infty} C_k\Phi_k(\mathbf{x})e^{-i(E_k - E_T)t}. \tag{3.22}$$

The coefficients C_k depend only on the initial conditions

$$C_k = \langle\Phi_k|\phi(t=0)\rangle. \tag{3.23}$$

The oscillatory time behavior of Eq. 3.22 is not of interest here; rather we wish to extract the ground state as above. This extraction can be achieved by substituting an *imaginary* time $\tau = it$ in Eqs. 3.21 and 3.22 so that the oscillatory time behavior becomes exponential, i.e.

$$-\frac{\partial\phi(\mathbf{x},\tau)}{\partial\tau} = (\mathcal{H} - E_T)\phi(\mathbf{x},\tau), \tag{3.24}$$

which has the solution

$$\phi(\mathbf{x},\tau) = \sum_{k=0}^{\infty} C_k\Phi_k(\mathbf{x})e^{-(E_k - E_T)\tau}. \tag{3.25}$$

For sufficiently large τ, only one eigenfunction contributes to ϕ, namely the one with the most negative eigenvalue — the ground state. Note, however, that even though we shall refer to the motion of particles, no real-time dynamics can be obtained from Eq. 3.24. Time-independence can be achieved by choosing E_T to be E_0, yielding

$$\phi(\mathbf{x},\tau) = C_0\Phi_0 + \sum_{k=1}^{\infty} C_k\Phi_k(\mathbf{x})e^{-(E_k - E_0)\tau}. \tag{3.26}$$

Thus, as $\tau \to \infty$, $\phi \to C_0\Phi_0$.

Let us now transform to the integral form as we did in Sec. 3.1.1 above and seek a solution of the form,

$$\phi(\mathbf{y}, \tau_2) = \int G(\mathbf{y}, \tau_2; \mathbf{x}, \tau_1)\phi(\mathbf{x}, \tau_1)\,\mathrm{d}\mathbf{x}. \tag{3.27}$$

An expression for $G(\mathbf{y}, \tau_2; \mathbf{x}, \tau_1)$ can be obtained by operating on both sides of Eq. 3.27 by $\mathcal{H}(\mathbf{y}) - E_T$, to yield

$$[\mathcal{H}(\mathbf{y}) - E_T]\phi(\mathbf{y}, \tau_2) = \int [\mathcal{H}(\mathbf{y}) - E_T]G(\mathbf{y}, \tau_2; \mathbf{x}, \tau_1)\phi(\mathbf{x}, \tau_1)\,\mathrm{d}\mathbf{x}, \tag{3.28}$$

and likewise, differentiating both sides of Eq. 3.27 with respect to τ_2, to yield

$$-\frac{\partial}{\partial \tau_2}\phi(\mathbf{y}, \tau_2) = -\int \frac{\partial G(\mathbf{y}, \tau_2; \mathbf{x}, \tau_1)}{\partial \tau_2}\phi(\mathbf{x}, \tau_1)\mathrm{d}\mathbf{x}\,. \tag{3.29}$$

Using Eq. 3.24, we can equate the right-hand sides of Eqs. 3.28 and 3.29, which after simplification yields,

$$-\frac{\partial G(\mathbf{y}, \tau_2; \mathbf{x}, \tau_1)}{\partial \tau_2} = (\mathcal{H} - E_T)G(\mathbf{y}, \tau_2; \mathbf{x}, \tau_1). \tag{3.30}$$

Thus the Green's function satisfies the same Schrödinger equation as the wave function ϕ. In addition, for $\tau_2 = \tau_1$ we have

$$\phi(\mathbf{y}, \tau_1) = \int G(\mathbf{y}, \tau_1, \mathbf{x}, \tau_1)\phi(\mathbf{x}, \tau_1)\mathrm{d}\mathbf{x}, \tag{3.31}$$

which implies that the Green's function satisfies

$$G(\mathbf{y}, \tau_1, \mathbf{x}, \tau_1) = \delta(\mathbf{x} - \mathbf{y}). \tag{3.32}$$

The structure of the time-dependent Green's function can be further elucidated by introducing the two times τ_1 and τ_2, as above, and writing the solution of Eq. 3.24 as

$$|\phi(\tau_2)\rangle = e^{-(\mathcal{H}-E_T)(\tau_2-\tau_1)}|\phi(\tau_1)\rangle. \tag{3.33}$$

(This form can be confirmed by substitution into Eq. 3.24.) The operator on the right-hand-side, $e^{-(\mathcal{H}-E_T)(\tau_2-\tau_1)}$, is the *time evolution* operator for $\mathcal{H} - E_T$. If we insert a complete set of position states between the operator and ϕ in Eq. 3.33, and multiply on the left by $\langle \mathbf{y}|$ we get

$$\phi(\mathbf{y}, \tau_2) = \int \langle \mathbf{y}| e^{-(\mathcal{H}-E_T)(\tau_2-\tau_1)} |\mathbf{x}\rangle \phi(\mathbf{x}, \tau_1)\, d\mathbf{x}. \tag{3.34}$$

By comparison with Eq. 3.27 we conclude that

$$G(\mathbf{y}, \tau_2; \mathbf{x}, \tau_1) = \langle \mathbf{y}| e^{-(\mathcal{H}-E_T)(\tau_2-\tau_1)} |\mathbf{x}\rangle. \tag{3.35}$$

This expression shows that the Green's function depends only on the difference $\tau_2 - \tau_1 \equiv \delta\tau$. Hence we can rewrite Eq. 3.27 as

$$\phi(\mathbf{y}, \tau + \delta\tau) = \int G(\mathbf{y}, \mathbf{x}; \delta\tau)\phi(\mathbf{x}, \tau)d\mathbf{x}. \tag{3.36}$$

As in the previous section, we have obtained an integral equation which can be solved iteratively.

The convergence of the series produced by iterating Eq. 3.36 can be studied by expanding $G(\mathbf{y}, \mathbf{x}; \delta\tau)$ in the eigenfunctions of $\mathcal{H}$. This expansion is achieved by inserting two complete sets of states into Eq. 3.35, and yields

$$G(\mathbf{y}, \mathbf{x}; \delta\tau) = \sum_{i=0}^{\infty} e^{-(E_i-E_T)\delta\tau} \Phi_i(\mathbf{y})\Phi_i(\mathbf{x}). \tag{3.37}$$

Now substitute this into Eq. 3.36 while also expanding $\phi(\mathbf{x}, 0) = \sum_k C_k\Phi_k$, to obtain the first iteration,

$$\begin{aligned}
\phi(\mathbf{y}, \delta\tau) &= \int \sum_{i=0}^{\infty} e^{-(E_i-E_T)\delta\tau} \Phi_i(\mathbf{y})\Phi_i(\mathbf{x}) \sum_{k=0}^{\infty} C_k\Phi_k(\mathbf{x})d\mathbf{x} \\
&= \sum_{k=0}^{\infty} C_k\Phi_k(\mathbf{y})e^{-(E_k-E_T)\delta\tau}.
\end{aligned} \tag{3.38}$$

This shows that the excited states decay exponentially fast. After n iterations we have,

$$\phi(\mathbf{y}, n\delta\tau) = \sum_{k=0}^{\infty} C_k \Phi_k(\mathbf{y}) e^{-(E_k - E_T)n\delta\tau}. \tag{3.39}$$

At large n, the lowest energy solution with a non-zero coefficient C_k will dominate the sum. Generally this will be the ground state, unless $\phi(\mathbf{x}, 0)$ is specifically chosen so that it is orthogonal to the ground state.

Although we have presented the time-dependent and time-independent formalisms separately, they are in fact intimately connected. The operators, $(\mathcal{H} - E_T)^{-1}$ and $e^{-\tau(\mathcal{H} - E_T)}$ are related by the Laplace transform,

$$\frac{1}{\mathcal{H} - E_T} = \int_0^{\infty} e^{-\tau(\mathcal{H} - E_T)} d\tau. \tag{3.40}$$

The Green's functions are the spatial representations of these two operators; hence they are related in the same way,

$$G(\mathbf{y}, \mathbf{x}) = \int_0^{\infty} G(\mathbf{y}, \mathbf{x}; \tau) d\tau. \tag{3.41}$$

The interpretation of the time-dependent Green's function is much the same as in the time-independent case. The function $G(\mathbf{y}, \mathbf{x}; \tau)$ represents the probability that a particle moves from $\mathbf{x}$ to $\mathbf{y}$ in an imaginary time τ. Starting from an initial ensemble of random walkers and propagating them iteratively in imaginary time according to the probabilities $G(\mathbf{y}, \mathbf{x}; \tau)$, the population density after a large number of generations will represent the ground state wave function. More specific details of how to use these Green's functions are presented in the next section.

3.2 Diffusion Monte Carlo

The central difficulty in Green's functions methods is obtaining the Green's function itself. In general, the exact Green's function is not known, for if it were the Monte

Carlo method would not be necessary! In this section we concentrate on a short-time approximation to obtain analytic (but approximate) Green's functions. This method is commonly called diffusion Monte Carlo (DMC) because of its underlying connection to a diffusion problem. The mapping of the imaginary time-dependent Schrödinger equation onto a classical diffusion equation is a connection noted at least as far back as Fermi.

Consider the imaginary-time electronic Schrödinger equation introduced in the previous section with the Hamiltonian now explicitly written out, i.e.

$$\frac{\partial \phi(\mathbf{x}, \tau)}{\partial \tau} = D \nabla^2 \phi(\mathbf{x}, \tau) + (E_T - V(\mathbf{x})) \, \phi(\mathbf{x}, \tau) \,, \qquad (3.42)$$

where D, the "diffusion" constant, is $\hbar^2/2m_e = 1/2$ in atomic units. As $\tau \to \infty$, asymptotically stationary behavior is obtained, and $\partial \phi / \partial \tau = 0$. In this limit Eq. 3.42 becomes the time-independent electronic Schrödinger equation. However, without the second term on the right hand side, Eq. 3.42 is the usual diffusion equation. Alternatively, ignoring the first term on the right hand side of Eq. 3.42 and retaining the second term results in a first-order rate equation or branching process whose "rate constant" is $(E_T - V)$. Both diffusion and rate processes can be simulated separately by the Monte Carlo method. It is therefore reasonable to expect that the entire equation could be simulated by a combined stochastic process consisting of diffusion plus branching. We have the interesting result that by transforming to imaginary time, a quantum mechanical problem can be simulated classically.

3.2.1   Short time approximation

To obtain an explicit expression for G, we can factor the time propagator, $e^{-(\mathcal{H} - E_T)\tau}$, into separate kinetic and potential energy Green's functions,

$$e^{-(\mathcal{T} + \mathcal{V} - E_T)\tau} \approx e^{-\mathcal{T}\tau} e^{-(\mathcal{V} - E_T)\tau} \equiv G_{\text{diff}} G_B. \qquad (3.43)$$

This factorization is only valid for small τ because $\mathcal{T}$ and $\mathcal{V}$ do not commute (see Exercise 3.2). The first correction term is of the form

$$G - G_{\text{diff}}G_B = \frac{1}{2}[\mathcal{V}, \mathcal{T}]\tau^2 + O(\tau^3).$$ (3.44)

We can identify G_{diff} as the Green's function of the classical diffusion equation, and G_B as the Green's function for the rate equation. The function G_{diff} satisfies the diffusion equation

$$-\frac{\partial G_{\text{diff}}(\mathbf{y}, \mathbf{x}; \tau)}{\partial \tau} = D\nabla^2 G_{\text{diff}}(\mathbf{y}, \mathbf{x}; \tau).$$ (3.45)

The solution to the one-dimensional diffusion equation is well known — it is a Gaussian spreading in time,

$$f(x, \tau) = (4\pi D\tau)^{-1/2} e^{-x^2/4D\tau},$$ (3.46)

where $f(x, \tau)$ is the density of diffusers. In analogy, the solution to Eq. 3.45 is a multidimensional Gaussian, symmetric in $\mathbf{x}$ and $\mathbf{y}$, spreading in τ,

$$G_{\text{diff}}(\mathbf{y}, \mathbf{x}; \tau) = (4\pi D\tau)^{-3N/2} e^{-(\mathbf{y}-\mathbf{x})^2/4D\tau}.$$ (3.47)

This part of the transition probability in the Monte Carlo iteration of Eq. 3.36 can be simulated by choosing random numbers from a Gaussian distribution whose variance is $2D\tau$.

The function G_B satisfies the rate equation $-\partial G_B/\partial \tau = (E_T - V)G_B$. The one-dimensional solution to the rate equation is

$$f(x, \tau) = e^{(V(x)-E_T)\tau}.$$ (3.48)

Therefore a symmetric, multidimensional form for G_B is

$$G_B(\mathbf{y}, \mathbf{x}; \tau) = e^{-(\frac{1}{2}[V(\mathbf{x})+V(\mathbf{y})]-E_T)\tau}.$$ (3.49)

This branching part of the Monte Carlo iteration of Eq. 3.36 can be simulated by the creation or destruction of particles (walkers) with probability G_B. Note that part of the approximation made in Eq. 3.43 is reflected in Eq. 3.49 because it ignores the path by which the walker moved from $\mathbf{x}$ to $\mathbf{y}$. Nevertheless, as $\tau \to 0$, $G_{\text{diff}}G_B$ converges to the exact Green's function.

Previously we showed that any initial distribution of walkers will converge to the ground state as $\tau \to \infty$, yet now we see that the DMC Green's function is accurate only as $\tau \to 0$. How does one satisfy both these requirements? We may do so by dividing the time τ into a large number of small time steps, $\delta\tau$, and iterating, i.e.

$$
\begin{aligned}
e^{-(\mathcal{H}-E_T)\tau} &= e^{-(\mathcal{H}-E_T)n\delta\tau} \\
&= e^{-(\mathcal{H}-E_T)\delta\tau} e^{-(\mathcal{H}-E_T)\delta\tau} e^{-(\mathcal{H}-E_T)\delta\tau} \cdots .
\end{aligned}
\tag{3.50}
$$

Long times are achieved by applying the short-time propagator many times. We are now able to describe a simple Monte Carlo simulation using the short-time approximate Green's function.

ALGORITHM 3.1 Simple diffusion Monte Carlo.

```
C
C Generate an initial ensemble of size Nwalkers from a previous VMC run.
C Propagate each walker and its offspring Nstep steps. GRAN is a random
C number generator for Gaussian variates, with a variance of 2*D*Tau.
C M is the current number of walkers, N is the number of electrons.
C
      Do Istep=1,Nstep
         M = Nwalkers
         Do k=1,M
C
C Initially sum potential for this walker over electrons.
C
            VX = POTENTIAL(X,k)
C
C Move each electron
C
            Do i=1,N
               Do j=1,3
                  X(i,j,k) = X(i,j,k) + GRAN(iseed)
               End Do
```

```
              End Do
C
C Compute potential and branching for new position.
C
              VY = POTENTIAL(X,k)
              GB = exp( -(0.5*(VY + VX) - ET) * Tau)
C
C Create a random integer whose average value is equal to the branching.
C
              MB = INT( GB + URAN(iseed) )
C
C Add MB - 1 copies to end of list. If MB = 0, then
C mark this configuration as dead.
C
              If (MB > 1) Then
                  Do n=1,MB - 1
                      Nwalkers = Nwalkers + 1
                      CALL COPY(X(k), X(Nwalkers))
                  End Do
              Else If (MB = 0) Then
                  Idead(k) = 1
              End If
          End Do (k loop)
C
C After all original M walkers are moved, remove the
C dead walkers and compact the list. First find the last alive walker.
C
          Do i=Nwalkers,1
              If (Idead(i) = 1) Then
                  Nwalkers = Nwalkers - 1
              End If
          End Do
          Do i=1,M
              If (Idead(i) = 1) Then
C
C Copy the last one on list
C
                  CALL COPY(X(Nwalker), X(i))
                  Nwalker = Nwalker - 1
              End If
          End Do
C
C Continue on to next iteration. After a sufficiently large number of
C generations the excited state contributions will have decayed. Then one may
C start gathering statistics on the ground state.
C
      End Do
      End
```

Let us illustrate this process with the hydrogen atom (see Fig. 3.2). For convenience we choose $E_T = -0.5$ hartrees, the exact ground-state energy of the atom. Select an arbitrary initial ensemble of walkers $\mathbf{X}^{(0)}$. According to the above discussion, the distribution of walkers at long times will converge to Φ_0. (Note that the density of walkers at a point is Φ_0 and *not* the electronic density given by $\Phi_0{}^2$.) The diffusion part of the Green's function, G_{diff}, leads to an asymptotically uniform distribution. The action of the branching term, through its **x** dependence, creates the shape of the Monte Carlo distribution.

In Fig. 3.2 the electron-proton Coulomb potential is indicated by the solid curve, and the dotted line indicates the trial energy level E_T. Where the two lines intersect is the radial distance r_0 at which $G_B = 1$. A walker with $r = r_0$ will not branch. Walkers with $r < r_0$ will have $G_B > 1$ ("births") and those with $r > r_0$ will have $G_B < 1$ ("deaths"). Therefore, the region $0 < r < r_0$ is a source of walkers while $r_0 < r < \infty$ is a sink for walkers. In fact, at $r = 0$ there is an infinite source. This problem is addressed in the next section. Thus the combined action of the kinetic (diffusion) and potential (branching) energies builds the structure of the wave function, shown as the heavy solid curve at the top of Fig. 3.2.

3.2.2   Importance sampling

Simulation of Eq. 3.42 by a random-walk-with-branching procedure as described in the previous section is inefficient. Among other reasons, the inefficiency is caused by the divergences in the branching rate constant $(E_T - V)$. These divergences lead to large fluctuations in the population and hence large statistical uncertainties in expectation values. These fluctuations can be greatly reduced by the Monte Carlo technique of *importance sampling* (cf. Ch. 1). In this procedure, one constructs an analytical trial function, Ψ, based on any available knowledge of Φ_0. Typically, Ψ is generated from standard methods, such as Hartree-Fock, etc. The trial function is then used to *bias* the random walk to produce the distribution $f(\mathbf{x}, \tau) \equiv \phi(\mathbf{x}, \tau)\Psi(\mathbf{x})$ rather than $\phi(\mathbf{x}, \tau)$. If we multiply Eq. 3.42 by Ψ and rewrite it in terms of $f(\mathbf{x}, \tau)$, we find,

$$\frac{\partial f(\mathbf{x}, \tau)}{\partial \tau} = D\nabla^2 f(\mathbf{x}, \tau) - D\nabla \cdot (f(\mathbf{x}, \tau)\mathbf{F}_Q(\mathbf{x})) + (E_T - E_L(\mathbf{x}))f(\mathbf{x}, \tau), \qquad (3.51)$$

Figure 3.2: Simple DMC for the hydrogen atom.

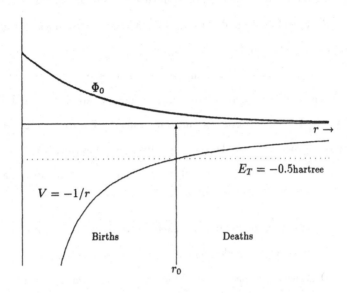

where $E_L \equiv \mathcal{H}\Psi/\Psi$ is the local energy, and $\mathbf{F}_Q \equiv \nabla \ln |\Psi|^2 = 2\nabla\Psi/\Psi$. The quantity $\mathbf{F}_Q(\mathbf{x})$ is a vector field, often called the *quantum force*. It is an effective velocity of the random walkers, directed away from regions where Ψ^2 is small.

Equation 3.51 (like Eq. 3.42) has terms that can be associated with diffusion and branching. The diffusion is modified, however, by a drift due to $\mathbf{F}_Q$. The drift term modifies the diffusion process precisely as particles undergoing Brownian motion are affected by an external field.

Another quantity appearing in Eq. 3.51 is the local energy, $E_L(\mathbf{x})$. The excess local energy $(E_T - E_L(\mathbf{x}))$ now appears as the branching term, rather than $(E_T - V(\mathbf{x}))$. The advantage of this change is that singularities in V are avoided. The local energy has both kinetic and potential energy terms and is much smoother than V alone. For the exact wave function the singularities in the kinetic and potential energies cancel, leaving $E_L(\mathbf{x}) = E_0$, a constant. In this limiting case, obtained by setting $\Psi = \Phi_0$ and $E_T = E_0$, the branching term drops out of Eq. 3.51 leaving an unchanging population of walkers sampled from $\Phi_0{}^2$. For approximate Ψ's, however, cancellation is not complete, and the population will change to compensate for the deficiences of Ψ. In particular, on average walkers will die out in regions where $\Psi > \Phi_0$ and give birth in regions where $\Psi < \Phi_0$. The largest branching fluctuations tend to occur where two particles come together and the potential diverges. However, this behavior can be minimized even for approximate Ψ's if the appropriate cusp conditions are met (see Ch. 5).

The local energy also serves as an energy estimator. In the limit that $\Psi = \Phi_0$, $E_L(\mathbf{x}) = E_0$, and only a single point is needed to evaluate the energy: there is no variance. Otherwise the magnitude of the variance is governed by the quality of Ψ. Most significantly, however, the expectation value of the local energy over the walk is E_0 *regardless* of the choice of Ψ. Importance sampling reduces the variance and increases the efficiency of the method, but does not influence the value of the energy.

Note that the VMC equation (Eq. 2.11) is identical to Eq. 3.51 without the branching term. It is easy to verify that without branching (as in VMC) $\phi(\mathbf{x}, \tau) \to \Psi$ and so $f(\mathbf{x}, \tau) \to \Psi^2$, as $\tau \to \infty$. Correspondingly, the average value of the local energy when branching is suppressed will be $\langle \Psi | \mathcal{H} | \Psi \rangle / \langle \Psi | \Psi \rangle$.

Importance sampling also changes the Green's function. The new Green's function, $\tilde{G}(\mathbf{y}, \mathbf{x}; \delta\tau)$, is a solution to Eq. 3.51 with the boundary condition $\tilde{G}(\mathbf{y}, \mathbf{x}; 0) = \delta(\mathbf{x} - \mathbf{y})$. The branching part can be obtained simply by replacing V with E_L in Eq. 3.49, namely

$$\tilde{G}_B(\mathbf{y}, \mathbf{x}; \delta\tau) = e^{-(\frac{1}{2}[E_L(\mathbf{x}) + E_L(\mathbf{y})] - E_T)\delta\tau}. \tag{3.52}$$

The modified "kinetic energy" operator in Eq. 3.51 is

$$\tilde{\mathfrak{J}} = -D\nabla^2 + D(\nabla \cdot \mathbf{F}) + D\mathbf{F} \cdot \nabla. \tag{3.53}$$

The new diffusion part of the Green's function is then

$$\tilde{G}_{\text{diff}}(\mathbf{y}, \mathbf{x}; \delta\tau) = e^{-\delta\tau\tilde{\mathfrak{J}}}. \tag{3.54}$$

By assuming that $\mathbf{F}_Q$ remains essentially constant over the move (which is increasingly true as $\delta\tau \to 0$), one can solve $\partial\tilde{G}_{\text{diff}}/\partial\tau = -\tilde{\mathfrak{J}}\tilde{G}_{\text{diff}}$ to obtain Eq. 2.22, i.e.

$$\tilde{G}_{\text{diff}}(\mathbf{y}, \mathbf{x}; \delta\tau) = (4\pi D\delta\tau)^{-3N/2} e^{-(\mathbf{x}-\mathbf{y}-D\delta\tau\mathbf{F}_Q(\mathbf{y}))^2/4D\delta\tau}. \tag{3.55}$$

This Green's function violates one of the properties specified earlier, namely

$$\tilde{G}_{\text{diff}}(\mathbf{y}, \mathbf{x}; \delta\tau) \neq \tilde{G}_{\text{diff}}(\mathbf{x}, \mathbf{y}; \delta\tau), \tag{3.56}$$

because $\tilde{\mathfrak{J}}$ is not Hermetian. It is therefore necessary to impose "detailed balance" as was done for VMC in Ch. 2 in order to guarantee equilibrium. Detailed balance is achieved by accepting the move of the walker from $\mathbf{x}$ to $\mathbf{y}$ with the Metropolis probability,

$$A(\mathbf{y}, \mathbf{x}; \delta\tau) \equiv \min(1, q(\mathbf{y}, \mathbf{x}; \delta\tau)), \tag{3.57}$$

where,

$$q(\mathbf{y}, \mathbf{x}; \delta\tau) \equiv \frac{|\Psi(\mathbf{y})|^2}{|\Psi(\mathbf{x})|^2} \frac{\tilde{G}(\mathbf{x}, \mathbf{y}; \delta\tau)}{\tilde{G}(\mathbf{y}, \mathbf{x}; \delta\tau)}. \tag{3.58}$$

This step insures that the distribution converges to $\Psi\Phi_0$ as $\delta\tau \to 0$. However, by rejecting some moves we have changed the effective step size, and so we must determine the proper time step to use in G_B. In Algorithm 3.2 below, each electron is moved and accepted or rejected individually. The effective time step is found by multiplying $\delta\tau$ by the ratio of accepted electron moves to the total attempted moves. With importance sampling the algorithm proceeds as follows:

ALGORITHM 3.2 Diffusion Monte Carlo with importance sampling.

```
C
C Choose an initial ensemble of size Nwalkers (see Algorithm 3.1). Before each
C walker begins, compute total energy and trial function at its starting point.
C As each electron is moved the energy and trial function is updated. GRAN is
C defined in Algorithm 3.1. Note that this algorithm is essentially the same as
C Algorithm 2.2, with the inclusion of branching.
C
      Do Istep=1,Nstep
         M = Nwalkers
         Do k=1,M
            Accept = 0.
            Do i=1,N
C
C Compute this electron's contribution to the trial function, PSIT, local
C energy, ELOCAL, and quantum force, FQ, for this walker. This is done in a
C subroutine which depends on the form of the trial function.
C
               CALL TRIALF(X,i,k,PSITX,ELOCALX,FQX)
C
C Move to a trial position and compute new quantities. NEWTRIALF only computes
C the change in the trial function due to moving electron i. See Appendix B
C for details.
C
               Do j=1,3
                  Y(j) = X(i,j,k) + D*Tau*FQX(j) + GRAN(iseed)
               End Do
               CALL NEWTRIALF(Y,i,k,PSITY,ELOCALY,FQY)
C Compute Metropolis acceptance probability.
C First compute log of the ratio of Green's functions.
C
               Q = 0
               Do k=1,3N
                  Q = Q + 1/2*(FQX + FQY)*[D*Tau/2*(FQX - FQY)
                                           - (Y(k) - X(i,k)]
               End Do
               A = min[ 1 , (PSITY/PSITX)**2*EXP(Q) ]
               If (A > URAN(iseed)) Then
                  CALL COPY(Y to X(i))
                  Accept = Accept + 1/N
               End If
            End Do
            Teff = Tau*Accept
C
C Compute branching factor and create or delete as in Algorithm 3.1.
C
            GB = exp( -(0.5*(ELOCALY + ELOCALX) - ET) * Teff)
            MB = INT( GB + URAN(iseed) )
                     .
                     .
                     .
            (see Algorithm 3.1)
                     .
                     .
                     .
         End Do
C
C After a sufficiently large number of generations, excited state
```

```
C contributions will have decayed.  Then one may start gathering
C statistics on the ground state distribution.
C
     End Do
     End
```

Returning to the example of hydrogen, in Fig. 3.3 we show the local energy for a good, though not exact, trial function. (We have used $\Psi = \exp(-x)\exp(-x^2/2)$ which has the wrong long range behavior but has the correct form at the nucleus, as discussed in Ch. 5.) Because of the drift term, the diffusion Green's function by itself produces asymptotically the Ψ^2 distribution. The branching term needs only to repair the relatively minor defects of the trial function to bring the distribution to $\Psi\Phi_0$. Note the much better behavior of E_L versus V in Fig. 3.2. Algorithm 3.2 is illustrative only. Many sampling refinements as well as reduction of the time step error can be achieved.[1] Most significantly, the time step bias can be greatly reduced by choosing a trial function that satisfies the *cusp* conditions discussed in greater detail in Ch. 5. These conditions ensure that the wave function has the correct limiting behavior at the singularities of the potential.

Another consideration is whether it is best to move and accept each electron individually as is done in Algorithm 3.2, or to move all the electrons at once, then accept the move as a whole. Both methods are in common use. Moving all electrons at once, then accepting the the total move, is attractive in part because this scheme is consistent with the exact Green's function algorithms to be described in Sec. 3.3. In this case, however, the effective time step must be computed separately, from a calculation of the total ratio of accepted to attempted moves. This ratio is then fixed throughout the walk. The disadvantage of moving all electrons at once is that the acceptance probability for a given time step is reduced considerably. Therefore in many-electron systems the one-electron at a time method is expected to be the most efficient.

Figure 3.3: Importance-sampled DMC for the hydrogen atom. For purposes of illustration the trial function is taken to be $\Psi = e^{-x}e^{-x^2/2}$.

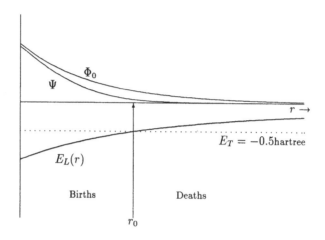

3.2.3   Estimating the ground state energy

The most readily available quantity from a DMC simulation is the ground state energy. Other properties, such as dipole and quadrupole moments, energy derivatives, and excited state energies will be discussed in Ch. 7. For now we will focus on two different estimators of the energy—the growth estimator and the local energy estimator.

¿From Eq. 3.39 we see that once all excited state contributions have died out, the remaining time dependence of the wave function is $e^{-(E_0-E_T)\tau}$. To obtain E_0 from this behavior, we can follow the growth of the population. Let the current total population of walkers (size of the ensemble) be denoted by $N(\tau)$. Clearly $f(\mathbf{x}, \tau)$ is not normalized, in fact the normalization *is* the total population at time τ, i.e.

$$N(\tau) = \int f(\mathbf{x}, \tau) d\mathbf{x}. \tag{3.59}$$

The population therefore changes as

$$N(\tau + \delta\tau) = e^{-(E_0 - E_T)\delta\tau} N(\tau). \tag{3.60}$$

For a sufficiently large sample of Monte Carlo points, the ground state energy can be estimated by the *growth* energy E_g, which can be obtained from the populations at an initial time τ_1 and at a later time τ_2 by rearranging Eq. 3.60 to solve for the ground-state energy, i.e.

$$E_g = E_T + \frac{1}{\tau_2 - \tau_1} \ln \frac{N(\tau_1)}{N(\tau_2)}. \tag{3.61}$$

The average of E_g taken over the course of the walk yields the estimate $E_0 = \langle E_g \rangle$. It is best to take the two times τ_1 and τ_2 as far apart as possible.[2] A difficulty with using E_g as an estimator is that the population can vary widely during a simulation, resulting in a large standard deviation of the mean.

The more commonly used estimator for the energy is E_L. It may seem odd that the exact energy can be estimated as the average value of the local energy of the trial function. In fact, in a VMC walk $\langle E_L \rangle_{\Psi^2} = E[\Psi]$ — i.e. it is explicitly trial function dependent. In a DMC walk, however, branching weights this average, resulting in

$$
\begin{aligned}
\langle E_L \rangle_f &= \int f_\infty(\mathbf{x}) E_L(\mathbf{x}) d\mathbf{x} / \int f(\mathbf{x}) d\mathbf{x} \\
&= \int \Phi_0(\mathbf{x}) \Psi(\mathbf{x}) \left[\frac{\mathcal{H}\Psi(\mathbf{x})}{\Psi(\mathbf{x})} \right] d\mathbf{x} / \int \Phi_0(\mathbf{x}) \Psi(\mathbf{x}) d\mathbf{x} \\
&= \int \Phi_0(\mathbf{x}) \mathcal{H}\Psi(\mathbf{x}) d\mathbf{x} / \int \Phi_0(\mathbf{x}) \Psi(\mathbf{x}) d\mathbf{x} \\
&= E_0.
\end{aligned}
\tag{3.62}
$$

The final step in Eq. 3.62 comes from the hermiticity of $\mathcal{H}$, which allows it to operate either forward (upon Ψ) or backward (upon Φ_0). Hence the DMC average of the local energy, i.e.

$$\langle E_L \rangle_f = \lim_{M \to \infty} \frac{1}{M} \sum_{k=1}^{M} E_L(\mathbf{X}_k), \tag{3.63}$$

will be exact to the extent that the distribution $f_\infty(\mathbf{x})$ is exact. However, because $\tilde{G}$ is only accurate to second order in $\delta\tau$, the distribution $f_\infty(\mathbf{x})$ and the resulting estimate of the energy will have a time step bias. One may eliminate this bias by evaluating $\langle E_L \rangle$ as a function of $\delta\tau$ and extrapolating to $\delta\tau = 0$.

3.2.4 "Pure" diffusion Monte Carlo

In the above discussion the action of G_B was taken to be the creation or annihilation of walkers. An alternative procedure[3, 4] which has advantages in certain situations, is to keep the value of G_B as a *weight* associated with each walker. In this situation the random walk is carried out without branching, leaving the population constant. Thus one has exactly the VMC walk, though with the addition of weights.

Let us develop the formalism for this "pure" diffusion Monte Carlo. If we begin with the integral equation for DMC with the initial distribution sampled from Ψ^2, i.e. $f(\mathbf{x}, 0) = \Psi^2$, then the first iteration gives us

$$
\begin{aligned}
f(\mathbf{y}, \delta\tau) &= \int G(\mathbf{y}, \mathbf{x}; \delta\tau) f(\mathbf{x}, 0) d\mathbf{x} \\
&= \int G_{\text{diff}}(\mathbf{y}, \mathbf{x}; \delta\tau) G_B(\mathbf{y}, \mathbf{x}; \delta\tau) \Psi^2(\mathbf{x}) d\mathbf{x}.
\end{aligned} \tag{3.64}
$$

Suppose that we iterate this equation n times, visiting the intermediate states, $\mathbf{x}_1$, ..., $\mathbf{x}_{n-1}$. The final distribution $f(\mathbf{x}_n, n\delta\tau)$ will be given by the integral over the intermediate states,

$$
\begin{aligned}
f(\mathbf{x}_n, n\delta\tau) &= \int G_{\text{diff}}(\mathbf{x}_n, \mathbf{x}_{n-1}; \delta\tau) G_B(\mathbf{x}_n, \mathbf{x}_{n-1}; \delta\tau) d\mathbf{x}_{n-1} \\
&\times \int G_{\text{diff}}(\mathbf{x}_{n-1}, \mathbf{x}_{n-2}; \delta\tau) G_B(\mathbf{x}_{n-1}, \mathbf{x}_{n-2}; \delta\tau) d\mathbf{x}_{n-2} \cdots \\
&\times \int G_{\text{diff}}(\mathbf{x}_1, \mathbf{x}_0; \delta\tau) G_B(\mathbf{x}_1, \mathbf{x}_0; \delta\tau) \Psi^2(\mathbf{x}_0) d\mathbf{x}_0.
\end{aligned} \tag{3.65}
$$

If we define the weight $w(\mathbf{x}_i)$ to be the value of the branching Green's function over the previous step, i.e. $w(\mathbf{x}_i) = G_B(\mathbf{x}_i, \mathbf{x}_{i-1}; \delta\tau)$, then we may rewrite Eq. 3.65 as,

$$
\begin{aligned}
f(\mathbf{x}_n, n\delta\tau) &= \int G_{\text{diff}}(\mathbf{x}_n, \mathbf{x}_{n-1}; \delta\tau) G_{\text{diff}}(\mathbf{x}_{n-1}, \mathbf{x}_{n-2}; \delta\tau) \cdots G_{\text{diff}}(\mathbf{x}_1, \mathbf{x}_0; \delta\tau) \\
&\quad \times \left\{ \prod_{i=1}^{n} w(\mathbf{x}_i) \right\} \Psi^2(\mathbf{x}_0) d\mathbf{x}_{n-1} d\mathbf{x}_{n-2} \cdots d\mathbf{x}_0.
\end{aligned}
\tag{3.66}
$$

Therefore the proper weight to be associated with each walker is the cumulative product of the branching weights. Let us define this cumulative weight for a walk ending at $\mathbf{x}_n$ as $W(\mathbf{x}_n)$,

$$
W(\mathbf{x}_n) = \prod_{i=1}^{n} w(\mathbf{x}_i).
\tag{3.67}
$$

Note that the random walk from $\mathbf{x}_0$ to $\mathbf{x}_n$ implied in Eq. 3.66 is just a VMC walk, propagated by G_{diff}, and hence generates an asymptotic distribution of Ψ^2. Expectation values, however, are computed using the weights. For example, for the energy we have,

$$
\begin{aligned}
\langle E_L \rangle &= \frac{\int \Psi^2(\mathbf{x}) W(\mathbf{x}) E_L(\mathbf{x}) d\mathbf{x}}{\int \Psi^2 W(\mathbf{x}) d\mathbf{x}} \\
&= \lim_{M \to \infty} \frac{\sum_{i=1}^{M} E_L(\mathbf{X}_i) W(\mathbf{X}_i)}{\sum_{i=1}^{M} W(\mathbf{X}_i)},
\end{aligned}
\tag{3.68}
$$

and this gives not the VMC energy but the DMC energy of Eq. 3.62.

As we will show in later chapters, the pure diffusion scheme has many uses because the branching weights may be manipulated directly. For the purpose of finding the ground state energy, however, it has a serious disadvantage. At long times one can see that the cumulative weight of a walker will either become very large (if the $w(\mathbf{x})$'s are on average greater than one) or vanish (if the weights are less than one on average). In an ensemble both behaviors may be present. Thus some walkers with nearly zero weight are kept together with others with a relatively large weight. (In a branching algorithm only the large weight walkers would survive and multiply.) This has two consequences. First, after a sufficiently large number of iterations the energy estimate will be dominated by a single walker, making the average meaningless. Second, the walk will sample space inefficiently, because a growing population of the walkers will effectively contribute nothing to the averages. Hence, if one's sole purpose is obtaining the ground state energy, the branching DMC algorithm is more efficient.

The pure DMC scheme does lead us to one improvement that can be readily made to the branching DMC algorithm.[5] Rather than determining the value of G_B and then finding an integer whose average value equals G_B, one can instead keep the weight as in pure DMC and branch whenever the cumulative weight is greater than or less than some chosen values. The integer part of the weight is then used to branch while the remainder is divided equally among the daughter walkers (if any). The combination of weights and branching allows one to create or destroy walkers less often (branching can be a time consuming part of the computer program) and the additional variance introduced by choosing an integer distributed as G_B is eliminated.

3.3   Exact Green's Function Methods

A disadvantage of the DMC method is that it contains a finite bias as long as $\delta\tau \neq 0$. DMC applications are critically dependent upon using small time steps and extrapolation to $\delta\tau = 0$. This presents difficulties because extrapolations are inherently ambiguous, and the extrapolation points are known only to within statistical errors.

It is both more accurate and more aesthetic to use methods without such bias. Much effort has gone into the development of methods that sample from the exact Green's function. For low dimensional systems, the Green's function can be expressed to arbitrary accuracy by expansions in simple orthogonal polynomials. This is done, e.g., in Gaussian wave-packet propagation methods, but this option is not feasible for molecular systems. However, for systems interacting via a bounded potential, i.e., for which the potential can be made negative everywhere by an appropriate shift of the zero of energy, one can write an exact Green's function in closed form. This is discussed further in Sec. 3.3.1. Again, molecular systems pose a problem because the Coulomb potential is *not* bounded. Hence Monte Carlo schemes have been developed specifically for the Coulomb case that do not require the exact Green's function but nevertheless use Monte Carlo to sample from it. In sampling from the exact Green's function one can begin from the time-independent development discussed in Sec. 3.3.2 or from the time-dependent approach, discussed in Sec. 3.3.3.

3.3.1 Green's function Monte Carlo for a bounded potential

The simplest exact Green's function method is one involving potentials that are strictly bounded from above. Consider the time-independent Schrödinger equation written in atomic units with an energy shift, i.e.

$$- \nabla^2 \Phi_0(\mathbf{x}) - 2(E_0 - V_{max})\Phi_0(\mathbf{x}) = -2(V(\mathbf{x}) - V_{\max})\Phi_0(\mathbf{x}). \qquad (3.69)$$

Let us assume that $V(\mathbf{x})$ has a finite upper bound, which is $V_{\max}$. In this case $V - V_{\max}$ will not be positive anywhere. We define $k^2 = -2(E_0 - V_{\max})$, and divide both sides of Eq. 3.69 by k^2 to obtain

$$[-\frac{1}{k^2}\nabla^2 + 1]\Phi_0(\mathbf{x}) = \left[\frac{V(\mathbf{x}) - V_{\max}}{E_0 - V_{\max}}\right]\Phi_0(\mathbf{x}). \qquad (3.70)$$

The Green's function for the operator on the left-hand side of this equation is known, namely

$$G_{\text{Bess}}(\mathbf{y}, \mathbf{x}) = (2\pi)^{-(\nu+1)}r^{-\nu}K_\nu(r), \qquad (3.71)$$

where $r = k\|\mathbf{x} - \mathbf{y}\|$ (i.e. the distance between the two points scaled by k), $\nu = 3N/2 - 1$, and K_ν is a Bessel function which is plotted in Fig. 3.4. Note that when ν takes on an integer value, K_ν is the modified Bessel function of the second kind;

when ν takes on a half-integer value, K_ν is the modified spherical Bessel function of the third kind.[6]

The wave function Φ_0 is obtained by iterating the integral equation

$$\Phi_0(\mathbf{y}) = \int G_{\text{Bess}}(\mathbf{y}, \mathbf{x}) \left[\frac{V(\mathbf{x}) - V_{\text{max}}}{E_0 - V_{\text{max}}} \right] \Phi_0(\mathbf{x}) d\mathbf{x}. \qquad (3.72)$$

Hence the total Green's function is $G_{\text{Bess}}(\mathbf{y}, \mathbf{x})(V(\mathbf{x}) - V_{\text{max}})/(E_0 - V_{\text{max}})$. Comparing this form with the DMC Green's function, Eq. 3.43, one sees a similar structure. G_{Bess} can be thought of as the "diffusion" part (i.e. the transition probability) while $(V(\mathbf{x}) - V_{\text{max}})/(E_0 - V_{\text{max}})$ acts as the branching part. The requirement of $V(\mathbf{x}) - V_{\text{max}} \leq 0$ insures that the branching is non-negative so that the Green's function can be interpreted as before. This means that to apply this approach to Coulomb systems one must "cut off" the repulsive electron-electron interaction at the chosen upper bound V_{max}. Such a cutoff will affect the energy, making this an approximate method for Coulomb systems. Some trial-and-error testing is needed to judge the impact of this approximation. In addition, the Coulomb repulsion term affects the average step size. For example, if we let $\langle r \rangle$ be the average value of the argument of the Bessel function, then the average step size is $\langle \|\mathbf{y} - \mathbf{x}\| \rangle = \langle r \rangle / k$. Therefore as $|E_0|$ and $|V_{\text{max}}|$ increases, the average step size decreases.

Importance sampling can be introduced by the multiplication of Eq. 3.72 by $\Psi(\mathbf{y})$ and multiplying and dividing the integrand by $\Psi(\mathbf{x})$. This yields the new integral equation

$$f(\mathbf{y}) = \int \frac{\Psi(\mathbf{y})}{\Psi(\mathbf{x})} G_{\text{Bess}}(\mathbf{y}, \mathbf{x}) \left[\frac{V(\mathbf{x}) - V_{\text{max}}}{E_0 - V_{\text{max}}} \right] f(\mathbf{x}) d\mathbf{x} \qquad (3.73)$$

where $f(\mathbf{x}) = \Phi_0(\mathbf{x}) \Psi(\mathbf{x})$ is the same quantity as in DMC.

How does one sample from this Green's function? Again we take G_{Bess} as the transition probability. Consider the case of a walker that moves from $\mathbf{x}$ to $\mathbf{y} \equiv \mathbf{x} + \Delta\mathbf{x}$. If the step $\Delta\mathbf{x}$ is expressed in spherical polar coordinates with $\mathbf{x}$ at the origin, one can

Figure 3.4: Plot of the Green's function of the Bessel method for $N = 1$–4. The order of the modified Bessel function is $\nu = 3N/2 - 1$.

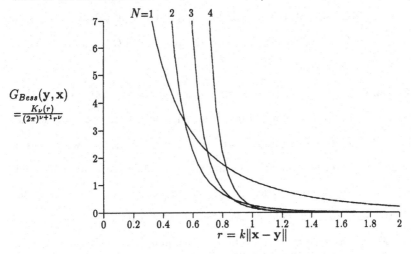

select $\Delta \mathbf{x}$ using Eq. 3.71 as a transition probability as follows. Since G_{Bess} does not depend on angle, begin by choosing a vector uniformly on the unit $(3N)$ sphere. It is not obvious, but this can be accomplished by generating a $3N$-component Gaussian random vector, and scaling the length of this vector to unity (i.e., dividing by its length). The probability distribution function for r is

$$F(r)dr = \frac{2^{\nu}\Gamma(\nu + \frac{1}{2})}{\Gamma(\frac{1}{2})\Gamma(2\nu + 1)} r^{\nu+1} K_{\nu}(r)dr, \tag{3.74}$$

where the coefficient gamma functions arise from normalization associated with the 3N-dimensional sphere. It can be shown[7] that r values sampled from Eq. 3.74 can be generated using $3N + 1$ uniformly distributed random numbers $\xi_0, \xi_1, \cdots, \xi_{3N}$ by generating two functions, namely

$$u = -\ln(\xi_1 \xi_2 \cdots \xi_{3N}) \tag{3.75}$$

and

$$v = (1 - \xi_0^{2/(3N-1)})^{1/2}. \tag{3.76}$$

Then $r = uv$. Together with the previously found point on the unit sphere this finishes the specification of $\Delta\mathbf{x}$. The remaining factors in the Green's function are used for weights and/or branching.

In the usual implementation, the branching is due to $G_B = (V(\mathbf{x}) - V_{\max})/(E - V_{\max})$, and is done before the move because it depends only on the potential at the initial point. An integer M is chosen as the integer part of $G_B + \xi$, where ξ is a uniform random number between zero and unity. The mean value of M is G_B, and so M copies of the walker are made. In addition, a weight $W = G_B/M$ may be associated with each copy to avoid statistical errors arising from the above integer rounding of M, as was suggested in the discussion of pure DMC. After the move, a Metropolis acceptance step, based on the ratio $\Psi(\mathbf{y})/\Psi(\mathbf{x})$, is used to bias the walk by the trial function. A great deal of care must be taken with this algorithm to assure regulation of the ensemble size, because the growth is based on $(V - V_{\max})/(E - V_{\max})$ even *with* importance sampling.[8]

ALGORITHM 3.3 Bessel function Monte Carlo.

```
C
C Choose an initial ensemble of size Nwalkers (see Algorithm 3.1). Then
C propagate each walker and its offspring Nstep steps. GRAN is defined as in
C Algorithm 3.1.
C
      Do Istep=1,Nstep
         M = Nwalkers
         Do i=1,M
C
C Compute trial function, PSIT, local energy, ELOCAL, and potential,
C V, for this walker and branch.
C
            CALL TRIALF(X(i),PSITX,ELOCALX,VX)
            GB = (VX - Vmax)/(E0 - Vmax)
               ... (Branch as in DMC) ...
C
C Move to a trial position and compute new quantities. First generate a random 3-N
```

```
C dimensional unit vector.
C
            Ylength = 0
            Do k=1,3*N
                Y(k) = GRAN(iseed)
                Ylength = Ylength + Y(k)**2
            End Do
            Ylength = Sqrt(Ylength)
C
C Now generate r, distributed as the Bessel function
C
            u = 1
            Do l=1,3*N
                u = u * URAN(iseed)
            End Do
            v = Sqrt(1 - (URAN(iseed)**(2/(3*N-1))))
            r = -Ln(u)*v
            Do k=1,3*N
                Y(k) = Y(k)*r/Ylength + X(i,k)
            End Do
            CALL TRIALF(Y,PSITY,ELOCALY,VY)
C
C Compute Metropolis acceptance probability
C
            A = min( 1 , PSITY/PSITX )
            If (A > URAN(iseed)) Then
                CALL COPY(Y to X(i))
            End If
        End Do
    End Do
    End
```

3.3.2 Domain Green's function Monte Carlo

The Bessel-function based method is particularly simple. Yet it has drawbacks — namely, the potential must be bounded above, and the branching factor is proportional to V. Here and in the following sections, we discuss an alternative method which uses intermediate Monte Carlo walks to *sample* the exact Green's function. In the literature both these types of method are often referred to as *Green's function Monte Carlo* (GFMC).

Of the approaches using intermediate walks, we begin with domain GFMC. It is a technique in which G is sampled without having been constructed explicitly. In the Bessel function method, the unboundedness of the Coulomb potential was a problem. In the domain approach we find the exact Green's function in separate finite spatial domains in which the potential is bounded, and then combine these domains to cover all space. Define a domain D_0 in $3N$ dimensional space as the direct product of 3D spheres centered on each electron. The radii of the individual spheres are chosen so that the potential is bounded everywhere within D_0 (i.e., by making sure that no sphere overlaps any other electron or nucleus). Choose a value U_0 greater than the maximum potential within D_0. The Green's function G_0 of the operator $\mathcal{H}_0 \equiv \mathcal{H} - V + U_0$ within D_0 is given by,

$$\mathcal{H}_0 G_0(\mathbf{z}, \mathbf{x}) = \delta(\mathbf{z} - \mathbf{x}) \tag{3.77}$$

within the domain, and zero elsewhere. Now we need to construct an equation relating the exact Green's function to G_0. This can be accomplished by multiplying Eq. 3.77 by $G(\mathbf{y}, \mathbf{z})$, and analogously multiplying $\mathcal{H}G(\mathbf{y}, \mathbf{z}) = \delta(\mathbf{y}-\mathbf{z})$ by $G_0(\mathbf{z}, \mathbf{x})$. Subtracting these two equations results in

$$G(\mathbf{y}, \mathbf{z})\delta(\mathbf{z} - \mathbf{x}) - G_0(\mathbf{z}, \mathbf{x})\delta(\mathbf{y} - \mathbf{z}) =$$
$$[\mathcal{H}_0 G_0(\mathbf{z}, \mathbf{x})]G(\mathbf{y}, \mathbf{z}) - [\mathcal{H}G(\mathbf{y}, \mathbf{z})]G_0(\mathbf{z}, \mathbf{x}). \tag{3.78}$$

By integrating over $\mathbf{z}$ and rearranging the terms, we obtain an iterative equation for G,

$$G(\mathbf{y}, \mathbf{x}) = G_0(\mathbf{y}, \mathbf{x}) + \int [\mathcal{H}_0 G_0(\mathbf{z}, \mathbf{x}) - G_0(\mathbf{z}, \mathbf{x})\mathcal{H}]G(\mathbf{y}, \mathbf{z})d\mathbf{z}. \tag{3.79}$$

We can write out $\mathcal{H}$ and $\mathcal{H}_0$ to reduce the right-hand side of Eq. 3.79 yielding,

$$G(\mathbf{y}, \mathbf{x}) = G_0(\mathbf{y}, \mathbf{x}) \quad + \int_{S_0} G(\mathbf{y}, \mathbf{z})[-\hat{n} \cdot \nabla G_0(\mathbf{z}, \mathbf{x})]d\mathbf{z}$$
$$+ \int_{D_0} G(\mathbf{y}, \mathbf{z})G_0(\mathbf{z}, \mathbf{x})[U_0 - V(\mathbf{x}) - E_T]d\mathbf{z}, \tag{3.80}$$

where we have introduced a constant energy offset E_T to $V(\mathbf{x})$. The purpose of this offset is, as in DMC, to keep the population close to constant on average. The second term on the right hand side is a surface integral, over the surface S_0 of the domain D_0. It comes from applying Green's theorem to the Laplacian terms of Eq. 3.79. (The operator $-\hat{n} \cdot \nabla$ is the gradient normal to the surface.) The third term on the right hand side is a volume integral over the difference between the approximate potential U_0 and the actual potential V. ¿From Eq. 3.77 and the definition of $\mathcal{H}_0$ we see that G_0 is the free particle Green's function (i.e., for diffusion) in the constant potential U_0. Hence,

$$G_0(\mathbf{y}, \mathbf{x}) = \int\limits_0^\infty G_{\text{diff}}(\mathbf{y}, \mathbf{x}, \tau) e^{-U_0 \tau} d\tau. \tag{3.81}$$

Here we have used the Laplace transform mentioned earlier to relate the time-independent to the time-dependent forms. The time step integration is accomplished by choosing random time steps during the walk.

Importance sampling can be introduced by multiplying Eq. 3.80 by $\Psi(\mathbf{y})$, and dividing by $\Psi(\mathbf{x})$. Defining the importance sampled Green's functions, $\tilde{G}(\mathbf{y}, \mathbf{x}) = \Psi(\mathbf{y}) G(\mathbf{y}, \mathbf{x}) / \Psi(\mathbf{x})$ and $\tilde{G}_0(\mathbf{y}, \mathbf{x}) = \Psi(\mathbf{y}) G_0(\mathbf{y}, \mathbf{x}) / \Psi(\mathbf{x})$ one has

$$\begin{aligned}
\tilde{G}(\mathbf{y}, \mathbf{x}) = \tilde{G}_0(\mathbf{y}, \mathbf{x}) \ &+ \ \int\limits_{S_0} \tilde{G}(\mathbf{y}, \mathbf{z})[-\hat{n} \cdot \nabla \tilde{G}_0(\mathbf{z}, \mathbf{x})] d\mathbf{z} \\
&+ \ \int\limits_{D_0} \tilde{G}(\mathbf{y}, \mathbf{z}) \tilde{G}_0(\mathbf{z}, \mathbf{x})[U_0 - V(\mathbf{x}) - E_T] d\mathbf{z}.
\end{aligned} \tag{3.82}$$

The exact Green's function is therefore obtained from the iterative solution of Eq. 3.80 or Eq. 3.82. This is accomplished by Monte Carlo in the same way as f or ϕ were generated earlier — by an iterative random walk. The first term of these equations is propagation of a walker directly from $\mathbf{x}$ to $\mathbf{y}$ by G_0 (or $\tilde{G}_0$). The second term corrects for the finite domain size by having G_0 generate intermediate walks to $\mathbf{z}$ on the surface of the domain. The final term is a correction to the potential U_0 by a walk, again generated by G_0, to an intermediate point $\mathbf{z}$ within the domain. The second and third terms thus generate new walkers in addition to those created by the first term. These walkers however are not yet part of the new generation. Rather, the intermediate walkers are added to the current generation's ensemble and must be iterated again by the full $G(\mathbf{y}, \mathbf{z})$. Only after all intermediate walkers and all their offspring have been processed and no new intermediate walkers have been created does one continue to the next generation. Although this sounds like a infinite process, the last two terms have an average probability of less than one for creating intermediate walkers.[9] Thus, the process will terminate, although these walkers are a source of considerable extra computation in this method.

The idea behind the domain method is simple: avoid the unbounded Coulomb potential by restricting movement in a domain chosen to exclude singularities. The implementation, however, is rather complicated. First a suitable domain must be chosen. As mentioned above, this involves determining suitable radii of the spheres around the individual electrons. One choice is to set these radii to be less than the smallest distance between electrons. In fact, to avoid overlap of the domains, frequently the choice is made to take $r_i < \frac{1}{2} \min(r_{ij})$. For many-electron systems, particularly involving heavy atoms, this will severely constrain the step size (because no walker is allowed to step outside the domain. Experience with the domain method[10] has shown that when applied to atomic and molecular systems, it converges about an order of magnitude more slowly than DMC. We include it here since it is nevertheless used for electronic structure calculations in a modified form. Moreover, it provides a simple picture of the iterative calculation of the Green's function. Below we give a simple implementation to illustrate some of the points of this method.

ALGORITHM 3.4 Domain Green's function Monte Carlo.

```
C
C Choose an initial ensemble of size Nwalkers,
C preferably from a previous DMC run.
C Unlike DMC here we keep two lists of walkers.
C Each walker in the first list is propagated by GO to obtain
```

```
C new direct walkers which are placed in the second list.
C The volume and surface terms generate walkers which are
C returned to the first list for further processing.
C The generation is finished only when no walkers remain in the
C first list.
C
        Do Istep=1,Nstep
           M = Nwalkers
           Do Until M = 0
C
C In this algorithm one processes the ensemble from last to first.
C
                i = M
                M = M - 1
C
C Generate direct configurations. Here we call a subroutine to
C choose a random time step and diffuse, drift and branch the
C particles by GO. It also keeps track of the number of walkers on the
C new list. See Algorithm 3.2 for details.
C
                CALL Propagate_by_GO(X(i) to Y(i))
C
C Generate configurations from the volume term. These are propagated
C by GO then branched with exp(U-V-ET) and returned to the first list.
C
                CALL Propagate_by_GO(X(i) to X(M+1))
                GBVol = exp( -(UO - VX - ET) * Tau)
                MBVol = INT( GBVol + URAN(iseed) )
            ...(Branch based on GBVol place MBVol copies at end of old list)...
                M = M + MBVol
C
C Generate configurations from the surface term. These are propagated
C by grad(GO) then returned to the first list.
C
                CALL Propagate_by_gradGO(X(i) to X(M+1))
                M = M + 1
C
C After all the walkers on the first list are gone, copy the second list
C to the first; return the total number of walkers and start the next generation.
C
                CALL COPY(Y to X return Nwalker)
                      ...(Compute the energy etc.)...
C
C Continue to the next iteration. Once the ensemble has equilibrated it
C will yield the ground state energy and distribution.
C
           End Do
        End Do
        End
```

The one step we did not discuss is the generation of walkers from the surface term. This process is rather complicated and is covered in detail in Ref. 9. In light of the above observations about the limited step size, we now turn to methods that do not impose the domain constraints.

3.3.3   Coulomb Green's function Monte Carlo

Here we discuss modifications of the above domain GFMC method enabling one to sample the exact Green's function for Coulomb systems without the domain constraint. This will make it possible to do away with the surface terms. This approach is in the time-dependent formalism.

To sample the exact time-dependent Green's function $G(\mathbf{y}, \mathbf{x}; \tau)$, one follows a procedure similar to that presented above for $G(\mathbf{y}, \mathbf{x})$. One proceeds by relating G to a known approximate Green's function G_0 by an integral equation, viz.,

$$G(\mathbf{y}, \mathbf{x}; \tau) = G_0(\mathbf{y}, \mathbf{x}; \tau) + \int \int_0^\tau G(\mathbf{y}, \mathbf{z}; \tau - t) K(\mathbf{z}, \mathbf{x}; t) \, d\mathbf{z} \, dt, \qquad (3.83)$$

where K is the kernel to be determined. Here t is the intermediate time (between zero and τ) corresponding to the move to $\mathbf{z}$. This equation is formally very similar to Eq. 3.79, though no domain restrictions have been placed upon it. To determine the kernel let us begin by writing the equation for the exact Green's function

$$\frac{\partial}{\partial t} G(\mathbf{y}, \mathbf{z}; \tau - t) = \mathcal{H}(\mathbf{z}) G(\mathbf{y}, \mathbf{z}; \tau - t), \qquad (3.84)$$

where an overall change in sign in the equation results from the $(-t)$ in the argument of G. If we then multiply both sides of Eq. 3.84 by $G_0(\mathbf{z}, \mathbf{x}; \tau)$ and integrate over t from 0 to τ as well as over the intermediate point $\mathbf{z}$, we obtain

$$\int \int_0^\tau G_0(\mathbf{z}, \mathbf{x}; t) \left[\frac{\partial}{\partial t} G(\mathbf{y}, \mathbf{z}; \tau - t) \right] d\mathbf{z} \, dt = \int \int_0^\tau G_0(\mathbf{z}, \mathbf{x}; \tau) [\mathcal{H}(\mathbf{z}) G(\mathbf{y}, \mathbf{z}; \tau - t)] d\mathbf{z} dt.$$
$$(3.85)$$

The kernel K in Eq. 3.83 is a *multiplicative* factor. Therefore we must reorganize Eq. 3.85 so that $\mathcal{H}$ and $\partial/\partial t$ operate on G_0 rather than the unknown function G. Because $\mathcal{H}$ is Hermetian it may act to the left on G_0. The derivative, $\partial/\partial t$, can be transferred from G to G_0 through integration by parts, so that the left-hand side of

Eq. 3.85, becomes

$$\int\int\limits_{0}^{\tau} G_0(\mathbf{z}, \mathbf{x}; t) \left[\frac{\partial}{\partial t} G(\mathbf{y}, \mathbf{z}; \tau - t)\right] d\mathbf{z}\, dt$$

$$= \int \left[G_0(\mathbf{z}, \mathbf{x}; t) G(\mathbf{y}, \mathbf{z}; \tau - t) |_0^\tau - \int\limits_0^\tau G(\mathbf{y}, \mathbf{z}; \tau - t)\frac{\partial}{\partial t} G_0(\mathbf{z}, \mathbf{x}; t)\, dt\right] d\mathbf{z}$$

$$= G_0(\mathbf{y}, \mathbf{x}; \tau) - G(\mathbf{y}, \mathbf{x}; \tau) - \int\int\limits_0^\tau G(\mathbf{y}, \mathbf{z}; \tau - t)\left[\frac{\partial}{\partial t} G_0(\mathbf{z}, \mathbf{x}; t)\right] d\mathbf{z}\, dt. \quad (3.86)$$

The final expression in Eq. 3.86 is obtained by employing the boundary condition $G_0(\mathbf{y}, \mathbf{x}; 0) = G(\mathbf{y}, \mathbf{x}; 0) = \delta(\mathbf{x} - \mathbf{y})$. Combining Eqs. 3.85 and 3.86 and rearranging terms, we obtain

$$G(\mathbf{y}, \mathbf{x}; \tau) = G_0(\mathbf{y}, \mathbf{x}; \tau) - \int\int\limits_0^\tau G(\mathbf{y}, \mathbf{z}; \tau - t)\left[\mathcal{H}(\mathbf{z}) + \frac{\partial}{\partial t}\right] G_0(\mathbf{z}, \mathbf{x}; t) d\mathbf{z} dt, \quad (3.87)$$

which identifies the kernel to be

$$K(\mathbf{z}, \mathbf{x}; t) = -\left[\mathcal{H}(\mathbf{z}) + \frac{\partial}{\partial t}\right] G_0(\mathbf{z}, \mathbf{x}; t). \quad (3.88)$$

Importance sampling is introduced just as in the domain GFMC method, by multiplying by $\Psi(\mathbf{y})$ and dividing by $\Psi(\mathbf{x})$, and replacing G by $\tilde{G}$ and G_0 by $\tilde{G}_0$ to obtain,

$$\tilde{G}(\mathbf{y}, \mathbf{x}; \tau) = \tilde{G}_0(\mathbf{y}, \mathbf{x}; \tau) + \int\int\limits_0^\tau \tilde{G}(\mathbf{y}, \mathbf{z}; \tau - t)\tilde{K}(\mathbf{z}, \mathbf{x}; t)\, d\mathbf{z}\, dt, \quad (3.89)$$

where

$$\tilde{K}(\mathbf{z}, \mathbf{x}; t) = -\frac{\Psi(\mathbf{z})}{\Psi(\mathbf{x})}\left[\mathcal{H}(\mathbf{z}) + \frac{\partial}{\partial t}\right] G_0(\mathbf{z}, \mathbf{x}; t). \quad (3.90)$$

At this point we have great freedom in choosing the form of G_0. An iterative solution of Eq. 3.87 will converge fastest if G_0 is close to G. Note that if $G_0 = G$, $K = 0$, and no iterations are necessary. Below we present two forms for G_0: one based upon a short-time-like approximation to the Feynman-Kac equation, and another based on the explicit sampling of the Coulomb singularity.

3.3.4   Feynman-Kac Coulomb correction

In the first of these two methods (which will be referred to as Coulomb 1), G_0 is taken to be a product of the free particle diffusion Green's function, G_{diff}, and a Coulomb correction term, i.e.

$$G_0(\mathbf{y}, \mathbf{x}; \tau) = G_{\text{diff}}(\mathbf{y}, \mathbf{x}; \tau) e^{-\upsilon(\mathbf{y}, \mathbf{x}; \tau)}, \qquad (3.91)$$

where υ is

$$\upsilon(\mathbf{y}, \mathbf{x}; \tau) = \int \int_0^\tau (V(\mathbf{z}) - E_T) \frac{G_{\text{diff}}(\mathbf{y}, \mathbf{z}; t) G_{\text{diff}}(\mathbf{z}, \mathbf{x}; \tau - t)}{G_{\text{diff}}(\mathbf{y}, \mathbf{x}; \tau)} d\mathbf{z}\, dt. \qquad (3.92)$$

This choice for υ is motivated by the Feynman-Kac expression

$$G(\mathbf{y}, \mathbf{x}; \tau) = G_{\text{diff}}(\mathbf{y}, \mathbf{x}; \tau) \left\langle e^{\left[-\int_0^\tau V(\mathbf{z}(t)) dt \right]} \right\rangle, \qquad (3.93)$$

where the average, $\langle \cdots \rangle$, is over all paths from $\mathbf{x}$ to $\mathbf{y}$. The approximate expression, Eq. 3.91, is formed by taking the average into the exponent — which becomes exact as $\tau \to 0$. Even for finite τ, this form is clearly an improvement over the domain method which ignores the Coulomb term altogether.

Substituting Eq. 3.91 for G_0 into Eq. 3.88 yields (after a great deal of manipulation) a simple form for the kernel[10]

$$K(\mathbf{z}, \mathbf{x}; \tau) = D \sum_i (\nabla_i \upsilon)^2 G_0(\mathbf{z}, \mathbf{x}; t). \qquad (3.94)$$

One important aspect of this expression is that $K \geq 0$, which satisfies the requirement that the transition probabilities and branching factors be non-negative.

The random walk is then the iterated sampling of Eq. 3.87 (or Eq. 3.89 if importance sampling is employed). Since we integrate over the intermediate time t, this method is formally equivalent to the time-*independent* domain method, where we introduced a fictitious time (cf. Eq. 3.81) so that the diffusion Green's function could be used to propagate the walkers, and we then integrated over time.

Once the time step has been chosen, direct configurations are generated from the first term on the right-hand side of Eq. 3.87 and intermediate configurations are generated by the kernel. Sampling the kernel is a difficult task in itself. The mathematics for doing so is given in Ref. 10. The direct propagation step also has added complexity: it cannot be carried out using G_0 directly because the normalization of G_0 is not known. Instead G_{diff} is used and the branching is multiplied by the ratio G_0/G_{diff} to correct for the difference. For programming purposes, the function $v(\mathbf{y}, \mathbf{x}; \tau)$, which is dependent only upon two independent variables, is calculated initially and fit by an appropriate polynomial. This algorithm is listed below and illustrated schematically in Fig. 3.5.

ALGORITHM 3.5 Coulomb 1 algorithm.

```
C
C Choose an initial ensemble of size Nwalkers,
C preferably from a previous DMC run.  This algorithm
C is similar to the domain GFMC approach, Algorithm 3.4.
C
      Do Istep=1,Nstep
         M = Nwalkers
         Do Until M = 0
C
C Processes ensemble from last to first.
C
            i = M
            M = M - 1
C
C Choose a random time step from exponential distribution.
C Then generate direct configurations and branch.
C
            CALL TRIALF(X(i),PSITX,ELOCALX,FQX)
            Tau = EXPRAN(iseed)
            CALL Propagate_by_GDiff(X(i) to Y(i) using Tau)
            CALL TRIALF(Y,PSITY,ELOCALY,FQY)
            GBDirect = Exp(ET*Tau)*PSITY/PSITX*G0(Y,X(i);Tau)/GDiff(Y,X(i);Tau)
C
C Generate intermediate walkers. These are generated from X(i) by
C branching alone.
C
            GBInt = Kernel(Y,X(i);Tau)*GBDirect*mu/G0(Y,X(i);Tau)
         ...(Branch based on GBInt; place copies at end of old list)...
            M = M + MBInt
C
C After all walkers on the first list are gone, copy the second list
C to the first and return the total number of walkers.
```

```
C
        CALL COPY(Y to X return Nwalker)
              ...(Compute the energy etc.)...
C
C Continue to next iteration. Once the ensemble has equilibrated it
C will yield the ground state energy and distribution.
C
        End Do
      End Do
      End
```

3.3.5   Explicit sampling of the Coulomb singularity

The second of the two methods (which will be referred to as Coulomb 2) is a modification of the domain GFMC method. For a given initial point $\mathbf{x}$, one chooses $U_0(\mathbf{x})$ to be a local upper bound to the potential as in the domain method. One then explicitly separates the potential into singular and non-singular parts,

$$-\frac{1}{r} = -V_s(r) + V_{ns}(r). \qquad (3.95)$$

The non-singular part is written as,

$$V_{ns}(r) = \frac{e^{-U_0 r}}{r} - \frac{1}{r}, \qquad (3.96)$$

and the singular term is the difference of V and V_{ns}, i.e. the Yukawa potential

$$V_s(r) = \frac{e^{-U_0 r}}{r}. \qquad (3.97)$$

If we substitute Eq. 3.95 into Eq. 3.87 and separate the result into singular and non-singular terms, we obtain

$$
\begin{aligned}
G(\mathbf{y}, \mathbf{x}; \delta\tau) \;=\;& P_1(\mathbf{y}, \mathbf{x}; \delta\tau) G_{\text{diff}}(\mathbf{y}, \mathbf{x}; \delta\tau) \\
&+ \int \int_0^{\delta\tau} G(\mathbf{y}, \mathbf{z}; \delta\tau - t) P_2(\mathbf{z}, \mathbf{x}; t) G_{\text{diff}}(\mathbf{z}, \mathbf{x}; t) d\mathbf{z} dt, \\
&+ \int \int_0^{\delta\tau} G(\mathbf{y}, \mathbf{z}; \delta\tau - t) P_3(\mathbf{z}, \mathbf{x}; t) \left\{ \frac{V_s(\mathbf{z}) G_{\text{diff}}(\mathbf{z}, \mathbf{x}; t)}{\int V_s(\mathbf{z}') G_{\text{diff}}(\mathbf{z}', \mathbf{x}; t) d\mathbf{z}'} \right\} d\mathbf{z} dt,
\end{aligned}
$$

$$(3.98)$$

Figure 3.5: Illustration of the Coulomb 1 algorithm, starting at point **x**.

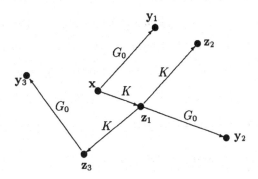

where P_1, P_2, and P_3 are probabilities of executing each term. These probabilities are

$$P_1(\mathbf{y}, \mathbf{x}; \delta\tau) = q(\mathbf{y}, \mathbf{x}; \delta\tau)e^{U_0(\mathbf{x})\delta\tau}, \qquad (3.99)$$

$$P_2(\mathbf{z}, \mathbf{x}; t) = q(\mathbf{z}, \mathbf{x}; t)[U_0(\mathbf{x}) - V_{ns}(\mathbf{x}) - E_T]e^{U_0(\mathbf{x})t}, \qquad (3.100)$$

and

$$P_3(\mathbf{z}, \mathbf{x}; t) = q(\mathbf{z}, \mathbf{x}; -t)\left[\int V_s(\mathbf{z}')G_{\mathrm{diff}}(\mathbf{z}', \mathbf{x}; t)d\mathbf{z}'\right]e^{U_0(\mathbf{x})t}, \qquad (3.101)$$

where q is the Metropolis weight,

$$q(\mathbf{y}, \mathbf{x}; t) = \frac{G_{\mathrm{diff}}(\mathbf{x}, \mathbf{y}, t)\Psi^2(\mathbf{y})}{G_{\mathrm{diff}}(\mathbf{y}, \mathbf{x}, t)\Psi^2(\mathbf{x})}. \qquad (3.102)$$

The above method removes the necessity for restricted domains and samples the Coulomb singularity exactly. As in the domain GFMC method the first term (chosen with probability P_1) creates direct walkers. The second term (chosen with probability P_2) corrects for the difference between U_0 and V_{ns}, and the third term (chosen with probability P_3) corrects for the difference between V_{ns} and V. The third term is large

only very near the singularity. The intermediate time t is chosen at random: from the

uniform distribution between 0 and $\delta\tau$ for the P_2 term and from the $t^{-1/2}$ distribution

for P_3. Full details are given in Ref. 11. The walker joins the new generation only

when it is propagated by the first term. All intermediate points are returned to the

current generation. Below we outline the Coulomb 2 algorithm.

ALGORITHM 3.6 Coulomb 2 algorithm.

```
C
C Choose an initial ensemble of size Nwalkers,
C preferably from a previous DMC run.
C
      Do Istep=1,Nstep
         M = Nwalkers
         Do Until M = 0
C
C Process ensemble from last to first.
C
            i = M
            M = M - 1
            CALL TRIALF(X(i),PSITX,ELOCALX,FQX)
C
C The diffusion time step is Tau1. Intermediate time steps Tau2 and Tau3
C are sampled as discussed in the text.
C
            Tau2 = URAN(iseed)*Tau1
            Tau3 = TSQRTRAN(iseed)
C
C Sample three possible steps Z1, Z2 and Z3.
C Calculate P1, P2 and P3 to determine which step is taken.
C VGint is a function which computes the integral of V singular and
C the Green's function (i.e. the integral in P3).
C
            CALL Propagate_by_GDiff(X(i) to Z1 using Tau1)
            CALL TRIALF(Z1,PSITZ1,ELOCALZ1,FQZ1)
            q = G(X(i),Z1,Tau1)*PSITX**2/[G(Z1,X(i),Tau1)*PSITZ1**2]
            P1 = q*Exp(-U0*Tau1)
            If (P1 > URAN(iseed)) Then
         ...(Branch as in DMC, add walkers to new generation Y)...
            Else
               CALL Propagate_by_GDiff(X(i) to Z2 using Tau1)
               CALL TRIALF(Z2,PSITZ2,ELOCALZ2,FQZ2)
               q = G(X(i),Z2,Tau1)*PSITX**2 / [ G(Z2,X(i),Tau1)*PSITZ2**2 ]
               P2 = q*Exp(-U0*Tau2)*{U0 - Vns(Z2) - ET}
               If (P2 > URAN(iseed)) Then
         ...(Branch as in DMC, add walkers to old generation X)...
                  M = M + Mbranching
               Else
                  CALL Propagate_by_GDiff(X(i) to Z3 using Tau1)
                  CALL TRIALF(Z3,PSITZ3,ELOCALZ3,FQZ3)
```

```
                    q = G(X(i),Z3,Tau1)*PSITX**2/[G(Z3,X(i),Tau1)*PSITZ3**2]
                    VG = VGint(Z3)
                    P3 = q*Exp(-U0*Tau3)*VG
                    GBs = Vs(Z3)
            ...(Branch as in DMC, based On Vs; add walkers to old generation X)...
C MBs is the integer whose average is GBs.
                    M = M + MBs
                End If
            End If
C
C After all walkers on the first list are gone, copy the second list back
C to the first. Return the total number of walkers and start the next generation.
C
            CALL COPY(Y to X return Nwalker)
                    ...(Compute the energy etc.)...
C
C Continue to the next iteration. Once the ensemble has equilibrated it
C will yield the ground state energy and distribution.
C
            End Do
        End Do
        End
```

3.4   Comparison of QMC Methods

We now compare the above methods to better understand their relative merits. From
the discussion of each method it is evident that the differences arise from differing
choices for G_0 and from whether the Green's function is known or is sampled. The
choices of G_0 and K are listed in Table 3.1, together with the resulting forms of direct
and indirect branching factors. ¿From this table we can see that the DMC and Bessel
methods are the simplest, and thus represent the fewest computations per step. The
GFMC methods which sample the Green's function have the additional computational burden of summing over all intermediate configurations. For the Coulomb 1
algorithm, computing K and processing the intermediate configurations roughly doubles the computation time per step. In the Coulomb 1 method, great care was taken
to use an accurate G_0 which will produce the smallest possible number of intermediate
configurations. Therefore the Domain and Coulomb 2 methods are even more costly,
since they are based on a much simpler G_0. Very few results exist in the literature
which allow direct comparison of the relative merits of these methods. In Table 3.2

Table 3.1: Comparison of the DMC, Bessel, and GMFC algorithms.

Method	G_0	K	Branching factor Direct	Branching factor Intermediate
Diffusion	$G_{\text{diff}}e^{-(E_L-E_T)\tau}$	none	$e^{-(E_L-E_T)\tau}$	none
Bessel	$G_{\text{Bess}}\frac{V-V_{\max}}{E_T-V_{\max}}$	none	$\frac{V-V_{\max}}{E_T-V_{\max}}$	none
Domain	$G_{\text{diff}}e^{-U_0\tau}$	$[(U-V-E_T)-\hat{\mathbf{n}}\cdot\nabla]G_0$	$e^{-U_0\tau}$	$U-V-E_T$
Coulomb 1	$G_{\text{diff}}e^{-\upsilon\tau}$	$[H+\frac{\partial}{\partial\tau}]G_0$	$e^{E_T\tau}\frac{G_0}{G_{\text{diff}}}$	$e^{E_T\tau}\frac{K}{G_{\text{diff}}}$
Coulomb 2	$G_{\text{diff}}e^{-U_0\tau}$	$[U_0-V_{ns}-E_T]G_0$	P_1	P_2 and P_3

we list for these methods the average time step size for several atoms and molecules. The average time step determines the speed at which the space of the electrons is sampled. The standard deviation has been shown[12] to depend on the time step $\delta\tau$ and is proportional to $\delta\tau^{-1/2}$. Hence in Monte Carlo, since σ is proportional to $1/M^2$, with M the number of samples, to achieve the same σ, M must be proportional to $1/\delta\tau$. In Table 3.2 only the smallest DMC time step is indicated. If one performed a time-step extrapolation, the "average" would be a factor of 3 to 10 times larger. Given this, DMC of all the methods uses the largest time steps for the heavier atoms. As the atomic charge increases, there is a marked decrease in the size of the time steps in all the methods. This decrease is especially marked for the Domain method due to the need to make the domains small enough for the potential to be bounded. For the Bessel method we have used the energy scaling, $k^2=-2(E_0-V_{\max})$, to estimate the time steps needed for Be and Ne based on the reported value for H_3^+ (Ref. 8). The Bessel method does not use a time step per se, but an effective time step can be

Table 3.2: Comparison of average time steps used in different QMC algorithms.

Method	H_2	H_3^+	He	LiH	Li_2	Be	Ne
DMC[10]	0.01		0.01	0:01	0.01		
DMC[1]					0.01	0.01	0.001
Domain[10]	0.053		0.022	0.0068	0.0037		
Domain[1]					0.0008	0.0006	0.00007
Coulomb 1[10]	0.55		0.43	0.087	0.043		
Coulomb 2[11]			0.005				
Bessel[8]		0.024				$(0.005)^a$	$(0.0006)^a$

(a) estimated values, see text.

determined from the Einstein equation,

$$\frac{1}{2}\frac{(\delta x)^2}{\delta \tau} = D \, , \tag{3.103}$$

where D is the diffusion constant from DMC. Therefore in the Bessel method the effective time step is inversely proportional to the energy, $E_0 - V_{max}$. The quantity V_{max} is an energy greater than the electron-electron Coulomb potential at an interelectronic distance r_0, where r_0 must be small enough that truncation of the potential at this value will not affect the total energy. (The r_0 value for He is reported[8] to be 0.1 bohr.) Therefore V_{max} is proportional to the number of electron pairs, i.e. to N^2, and inversely proportional to r_0. Given that the $V_{max} = -15.0$ hartrees used for H_3^+ was reported[8] to yield an effective time step of 0.024 hartrees^{-1}, we have estimated the values for Be and Ne. Of all the GFMC methods the Coulomb 1 algorithm seems to have the best time-step behavior. This is a result of the high accuracy of the kernel.

In Tables 3.3 and 3.4 we list a number of representative results for the methods. One sees that in spite of the theoretical differences between the methods, each is capable of high accuracy for a wide range of systems. The relative errors however do not reflect on the method used because each result represents a different amount

Table 3.3: Energy-related properties of atoms and molecules computed by DMC.

Property	System	QMC	Experiment	Refs.
Singlet-triplet splitting (kcal/mole)	CH_2	8.9(2.2)	9.05(6)	22
Electron affinity (eV)	F	3.45(11)	3.399	23
Barrier height (kcal/mole)	$H + H_2$	9.61(1)	9.65(8)	21
	$F + H_2$	3.2(1.3)	—	20
Binding energy (kcal/mole)	N_2	217.1(2.4)	228.4	18
Reaction energy (kcal/mole)	$F + H_2 \rightarrow FH + H$	29.7(1.6)	31.7(2)	20

of computation time. In addition, all the results except for those obtained with the Coulomb 1 method employ the fixed-node approximation discussed in Ch. 4. Comparing the relative achieved precision of the Domain and DMC methods among the most recent results (which we assume represent the state of the art), we see a factor of 20 in statistical error for N_2 and 2.5 for LiH, in both cases favoring DMC. For Li_2 and Be, using a more advanced form of DMC[1] than presented in Sec. 3.2, the ratio of errors between Domain and DMC was 30 and 167, respectively. However, as these algorithms are improved and their computational speed increased, one can expect further error reduction for all the methods.

Table 3.4: Total energies of atoms and molecules computed by QMC.

System	Method	QMC Energy[a]	%CE	Ref.
Li	DMC	$-7.4784(1)$	100.0(5)	13
Li	Domain	$-7.4790(20)$	103(4)	14
LiH	DMC	$-8.0700(4)$	99.5(5)	15
LiH	Domain	$-8.0699(10)$	100(1)	14
LiH	Coulomb 1	$-8.0710(10)$	100(1)	16
Li_2	DMC	$-14.9890(2)$	98.2(8)	1
Li_2	Domain	$-14.9910(60)$	98(5)	14
Li_2	Coulomb 1	$-14.9940(20)$	100(2)	16
Be	DMC	$-14.66717(3)$	100.5(3)	1
Be	Domain	$-14.6550(50)$	88(5)	14
BeH	Domain	$-15.2470(40)$	92(5)	14
B	Domain	$-24.6430(40)$	92(3)	14
B_2	Domain	$-49.9910(350)$	86(11)	14
C	Domain	$-37.8280(120)$	90(8)	14
CH	Domain	$-38.4650(150)$	94(8)	14
CH_4	DMC	$-40.5063(22)$	97.3(7)	17
N	DMC	$-54.5765(12)$	93.1(6)	18
N_2	DMC	$-109.4835(37)$	90.5(7)	18
N_2	Domain	$-109.5170(790)$	96(14)	14
O	Domain	$-75.0590(270)$	98(11)	14
HO	Domain	$-75.7240(260)$	97(8)	14
H_2O	DMC	$-76.4024(26)$	90.5(7)	19
H_2O	Coulomb 1	$-76.4300(200)$	100(5)	16
F	DMC	$-99.7167(13)$	95.0(4)	20

(a) All energies (in hartrees) were obtained in the fixed node approximation except for Coulomb 1.

Exercises

1. Here we will explore further the Green's function for a particle in a one-dimensional box. Implement a random walk procedure using Eq. 3.13 by the following procedure:

   (a) Choose an initial point x uniformly within the box.

   (b) Compute the area A under $G(y, x)$. (This is the sum of the two triangular regions.)

   (c) Generate a random integer M which has an average value of $E_T A$.

   (d) Choose M points from the normalized triangular-shaped distribution G/A.

   (e) By performing (b) through (d) measure the shape of the resulting distribution for Φ_0.

2. Derive Eq. 3.37. How do you deal with the exponential of an operator?

3. Show that Eq. 3.43 is an approximate relationship and that Eq. 3.44 is the first correction term.

4. Derive Eq. 3.51 from Eq. 3.42. (Hint: first obtain Eq. 3.42 from Eq. 3.51.)

5. Write and execute a simple diffusion Monte Carlo program for H_2 without importance sampling.

6. Modify the program from Exercise 5 to include importance sampling. Note that the local kinetic energy of a hydrogenic orbital is simply the potential subtracted from the total energy, which is

$$E_n = -\frac{\zeta^2}{2n^2}, \tag{3.104}$$

where n is the principle quantum number and ζ is the exponent in the orbital. Use this program to treat other two-electron singlet state atoms and diatomics.

7. Write and execute a computer program for the Bessel GFMC algorithm for H_2. Compare the results with the DMC results obtained in the previous exercises.

Suggestions for Further Reading

1. A. C. Hurley, *Introduction to the Electronic Theory of Small Molecules* (Academic, New York, 1976).

2. A. C. Hurley, *Electron Correlation in Small Molecules* (Academic, New York, 1976).

3. For a comprehensive listing of Green's function quantum Monte Carlo references up to 1990 see, W. A. Lester, Jr. and B. L. Hammond, "Quantum Monte Carlo for the Electronic Structure of Atoms and Molecules," *Annual Reviews of Physical Chemistry* **41**, 283-311 (1990).

4. R. Grimm and R. G. Storer, "A New Method for the Numerical Solution of the Schrödinger Equation," *Journal of Computational Physics* **4**, 230-49 (1969).

5. M. H. Kalos, "Optimization and the Many-Fermion Problem," in *Monte Carlo Methods in Quantum Problems*, 19-31 (D. Reidel, Dordrecht, 1984).

6. D. M. Ceperley, "A Review of Quantum Monte Carlo Methods and Results for Coulombic Systems," in *Monte Carlo Methods in Quantum Problems*, 47-57 (D. Reidel, Dordrecht, 1984).

7. J. W. Moskowitz and K. E. Schmidt, "Can Monte Carlo Methods Achieve Chemical Accuracy?," in *Monte Carlo Methods in Quantum Problems*, 59-70, (D.

Reidel, Dordrecht, 1984).

8. K. E. Schmidt and M. H. Kalos, "Few- and Many-Fermion Problems," in *Monte Carlo Methods in Statistical Physics II*, 125-44 (Springer-Verlag, New York, 1984).

9. M. H. Kalos and P. A. Whitlock, *Monte Carlo Methods Volume 1: Basics* (Wiley, New York, 1986).

10. B. H. Wells, "Green's Function Monte Carlo Methods," in *Methods in Computational Chemistry 1*, 311-50 (Plenum, New York, 1987).

11. M. Caffarel, "Stochastic Methods in Quantum Mechanics," in *Numerical Determination of the Electronic Structure of Atoms, Diatomic and Polyatomic Molecules*, 85-105 (Kluwer Academic Publishers, 1989).

12. P. J. Reynolds, J. Tobochnik and H. Gould, "Diffusion Quantum Monte Carlo," *Computers in Physics* Nov/Dec, 882-88 (1990).

13. M. A. Lee and K. E. Schmidt, "Green's Function Monte Carlo," *Computers in Physics* Mar/Apr, 192-97 (1992).

14. B. L. Hammond, M. M. Soto, R. N. Barnett and W. A. Lester, Jr., "On Quantum Monte Carlo for the Electronic Structure of Molecules," *Journal of Molecular Structure (Theochem)* **234**, 525-38 (1991).

References

1. C. J. Umrigar, M. P. Nightingale and K. J. Runge, "A Diffusion Monte Carlo Algorithm with Very Small Time-step Errors," *Journal of Chemical Physics* **99**, 2865-90 (1993).

2. D. M. Ceperley and M. H. Kalos, "Quantum Many-Body Problems," in *Monte Carlo Methods in Statistical Physics*, 145-94 (Springer-Verlag, New York, 1979).

3. M. Caffarel and P. Claverie, "Development of a Pure Diffusion Quantum Monte Carlo Method Using a Full Generalized Feynman-Kac Formula. I. Formalism," *Journal of Chemical Physics* **88**, 1088-99 (1988).

4. M. Caffarel and P. Claverie, "Development of a Pure Diffusion Quantum Monte Carlo Method Using a Full Generalized Feynman-Kac Formula. II. Applications to Simple Systems," *Journal of Chemical Physics* **88**, 1100-9 (1988).

5. R. N. Barnett, P. J. Reynolds and W. A. Lester, Jr., "Monte Carlo Algorithms for Expectation Values of Coordinate Operators," *Journal of Computational Physics* **96**, 258-76 (1991).

6. M. Abramowitz and I. Stegun, eds., *Handbook of Mathematical Functions*, (Dover, New York, 1972). The modified Bessel functions are discussed on p. 374 and the modified spherical Bessel function are given on p. 444.

7. M. H. Kalos, "Monte Carlo Calculations of the Ground State of Three- and Four-Body Nuclei," *Physical Review* **128**, 1791-95 (1962).

8. J. B. Anderson, "Simplified Sampling in Quantum Monte Carlo: Application to H_3^+," *Journal of Chemical Physics* **86**, 2839-43 (1987).

9. M. H. Kalos, D. Levesque and L. Verlet, "Helium at Zero Temperature with Hard-Sphere and Other Forces," *Physical Review A* **9**, 2178-95 (1974).

10. D. Ceperley, "The Simulation of Quantum Systems with Random Walks: A New Algorithm for Charged Systems," *Journal of Computational Physics* **51**, 404-22 (1983).

11. D. W. Skinner, J. W. Moskowitz, M. A. Lee, P. A. Whitlock, and K. E. Schmidt, "The Solution of the Schrödinger Equation in Imaginary Time by Green's Function Monte Carlo. The Rigorous Sampling of the Attractive Coulomb Singularity," *Journal of Chemical Physics* **83**, 4668-72 (1985).

12. S. M. Rothstein and J. Vrbik, "Statistical Error of Diffusion Monte Carlo," *Journal of Computational Physics* **74**, 127-42 (1988).

13. R. N. Barnett, P. J. Reynolds and W. A. Lester, Jr., "Monte Carlo Determination of the Oscillator Strength and Excited State Lifetime for the Li 2^2S $\rightarrow$ 2^2P Transition," *International Journal of Quantum Chemistry* **42**, 837-47 (1992).

14. R. P. Subramaniam, M. A. Lee, K. E. Schmidt and J. W. Moskowitz, "Quantum Simulation of the Electronic Structure of Diatomic Molecules," *Journal of Chemical Physics* **97**, 2600-9 (1988).

15. R. N. Barnett, P. J. Reynolds and W. A. Lester, Jr., "Is Quantum Monte Carlo Competitive: Lithium Hydride Test Case," *Journal of Physical Chemistry* **91**, 2004-5 (1987).

16. D. Ceperley and B. J. Alder, "Quantum Monte Carlo for Molecules: Green's Function and Nodal Release," *Journal of Chemical Physics* **81**, 5833-44 (1984).

17. D. R. Garmer and J. B. Anderson, "Quantum Chemistry by Random Walk: Methane," *Journal of Chemical Physics* **86**, 4025-29 (1987).

18. P. J. Reynolds, R. N. Barnett, B. L. Hammond and W. A. Lester, Jr., "Molecular Physics and Chemistry Applications of Quantum Monte Carlo," *Journal of Statistical Physics* **43**, 1017-26 (1986).

19. P. J. Reynolds, D. M. Ceperley, B. J. Alder and W. A. Lester, Jr., "Fixed-Node Quantum Monte Carlo for Molecules," *Journal of Chemical Physics* **77**, 5593-603 (1982).

20. D. R. Garmer and J. B. Anderson, "Quantum Chemistry by Random Walk: Application to the Potential Energy Surface for $F + H_2 \rightarrow HF + H$," *Journal of Chemical Physics* **86**, 7237-39 (1987).

21. D. L. Diedrich and J. B. Anderson, "An Accurate Quantum Monte Carlo Calculation of the Barrier Height for the Reaction $H + H_2 \rightarrow H_2 + H$," *Science* **258**, 786-88 (1992).

22. P. J. Reynolds, M. Dupuis and W. A. Lester, Jr., "Quantum Monte Carlo Calculation of the Singlet-Triplet Splitting in Methylene," *Journal of Chemical Physics* **82**, 1983-90 (1985).

23. R. N. Barnett, P. J. Reynolds and W. A. Lester, Jr., "$H + H_2$ Reaction Barrier: A Fixed-Node Quantum Monte Carlo Study," *Journal of Chemical Physics* **82**, 2700-7 (1985).

Chapter 4

Treating Fermions

A fundamental requirement of Fermi statistics is that an electronic wave function be antisymmetric with respect to exchange of any two electrons. This requirement means that

$$\Phi^F(\cdots \mathbf{x}_i \cdots \mathbf{x}_j \cdots) = -\Phi^F(\cdots \mathbf{x}_j \cdots \mathbf{x}_i \cdots). \qquad (4.1)$$

On the other hand, the mathematical ground state for the electronic Hamiltonian is *symmetric*, i.e., it obeys Bose statistics. The Fermi ground state is in fact the first fully antisymmetric excited state of the Hamiltonian. Later chapters will specifically address the issue of obtaining higher excited states. The present discussion will focus rather on determining and imposing the appropriate boundary conditions for obtaining only the Fermion states and, in particular, the Fermi ground state.

The central difficulty in the QMC simulation of Fermions by the methods we have described thus far, is that the exact wave function is represented by a *density* of random walkers, and that this density analogy breaks down for antisymmetric states. This failure results because the wave function has both negative and positive regions. To simulate such systems we must find a method of maintaining a positive density of walkers everywhere, while still enforcing the antisymmetry condition.

In this chapter we present methods that impose the antisymmetry. We group

131

them into two classes: *nodal* methods and *interacting walker* methods. Nodal methods are based upon the simplest, and historically the first method used to impose antisymmetry: the fixed-node method. This method is approximate because it relies on the nodes of an approximate trial function to determine the nodes of the QMC result. Because of this approximation, methods have been developed to release the nodal constraints and to find the exact wave function. The resulting method, unfortunately, provides only a *transient* estimate, i.e., the solution is not stable in imaginary time. Interacting ensemble methods seek to stabilize the separation of positive and negative populations, and to build antisymmetry by non-local interactions within the ensemble.

4.1   Nodal Methods

4.1.1   Fixed-node approximation

Let us begin by recalling the importance-sampled imaginary-time Schrödinger equation, Eq. 3.51,

$$-\frac{\partial f(\mathbf{x}, \tau)}{\partial \tau} = -D\nabla^2 f(\mathbf{x}, \tau) + (E_L(\mathbf{x}) - E_T)f(\mathbf{x}, \tau) + D\nabla \cdot (f\mathbf{F}_Q(\mathbf{x})), \qquad (4.2)$$

where the density function $f(\mathbf{x}, \tau) = \phi(\mathbf{x}, \tau)\Psi(\mathbf{x})$. In Ch. 3 it was shown that at large τ, f will converge to $f_\infty(\mathbf{x}) = \Phi_0(\mathbf{x})\Psi(\mathbf{x})$. Because $f(\mathbf{x}, \tau)$ is a density, we have the implicit boundary condition $\phi\Psi \geq 0$ for Eq. 4.2. That is, $f(\mathbf{x}, \tau)$ can be represented by a density of walkers only as long as Ψ and ϕ change sign together and thus share the same nodes (see Fig. 4.1).

This boundary condition can be enforced during a simulation by fixing the nodes[1] of $\phi(\mathbf{x}, \tau)$ to be those of Ψ, and rejecting the move of a walker that attempts to cross a node of Ψ. We shall denote the asymptotic solution of Eq. 4.2 with this constraint

Figure 4.1: Fixed-node approximation for a simple one-dimensional wave function. The nodes of Φ_0^{fn} are identical to those of the trial function Ψ. Errors in the nodes of Ψ result in deviations in Φ_0^{fn} from the exact wave function Φ_0^F.

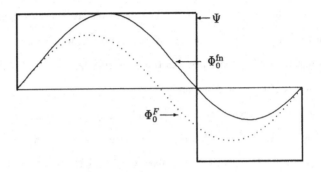

by

$$f_\infty(\mathbf{x}) = \Phi_0^{fn}(\mathbf{x})\Psi(\mathbf{x}). \tag{4.3}$$

The fixed-node solution, Φ_0^{fn}, is an approximation to the exact Fermi ground state Φ_0^F. If the nodes of Ψ were exact, then the function Φ_0^{fn} would also be exact. Although it is unlikely that one can find the exact nodes (other than in one dimension), it is useful to test the quality of this approximation. We primarily are interested in the fixed-node energy, E^{fn}. We can use a perturbation theory argument to see how E^{fn} depends upon the defect in the nodes. Define

$$\Phi_0^F = \Phi_0^{fn} + \lambda\Phi_1^{fn} + \lambda^2\Phi_2^{fn} + \cdots, \tag{4.4}$$

where λ is the nodal "perturbation" and Φ_1^{fn}, Φ_2^{fn}, $\cdots$ are the first-, second-order, etc., wave functions orthogonal to Φ_0^{fn}. The zeroth-order energy is the fixed-node energy.

The first-order energy vanishes, since

$$\begin{aligned} E^{(1)} &= \langle \lambda \Phi_1^{\text{fn}} | \mathcal{H} | \Phi_0^{\text{fn}} \rangle + \langle \Phi_0^{\text{fn}} | \mathcal{H} | \lambda \Phi_1^{\text{fn}} \rangle \\ &= 2\lambda E^{\text{fn}} \langle \Phi_1^{\text{fn}} | \Phi_0^{\text{fn}} \rangle = 0, \end{aligned} \qquad (4.5)$$

due to the orthogonality condition. The first nonzero energy correction is second-order in the "nodal defect". Because of this property, we expect that the fixed-node approximation is fairly insensitive to the quality of Ψ, and can be accurate even for simple choices of the trial function.

It also can be shown that the fixed-node energy is a variational upper bound to E_0. To show this, let us divide the $3N$-dimensional space of the N electrons into the nodal volumes ν_i. In each of these separate volumes there is a ground-state solution Φ_i with energy ϵ_i, which satisfies the Schrödinger equation inside that volume and is zero elsewhere. As seen in Ch. 3, the application of QMC separately to each of these volumes will reproduce these eigenfunctions and eigenvalues.

The key to the variational argument is that each ϵ_i is itself an upper bound to the ground state energy. This property can be seen by constructing a global antisymmetric wave function by permutations of the solution from volume ν_i, i.e.,

$$\hat{\Phi}^{(i)}(\mathbf{x}) = \sum_P (-1)^P \Phi_i(\mathcal{P}\mathbf{x}), \qquad (4.6)$$

where $\mathcal{P}$ is the permutation operator on the electrons, P is its eigenvalue, and $(-1)^P$ is $+1$ for even permutations and -1 for odd ones. Each function $\hat{\Phi}^{(i)}$ covers all space with replicas of Φ_i, and therefore has energy expectation value ϵ_i. From the variational principle. ϵ_i thus must be a variational upper bound to E_0^F. If all nodal volumes were equivalent, then $\hat{\Phi}^{(1)} = \hat{\Phi}^{(2)} = \cdots \hat{\Phi}^{(n)}$ (as is true of the exact ground state), and the QMC energy would be $E^{\text{fn}} = \epsilon_i$. If all volumes are not equivalent (as is the case, e.g., for higher excited states), then the QMC population in those volumes with higher energy will shrink relative to the population in lower energy volumes. Ultimately, only the minimum energy volume will be populated, and $E^{fn} = \epsilon_{min}$. In either case, the QMC fixed-node energy remains a rigorous upper bound to the exact ground state Fermi energy. Implementation of the fixed-node approximation requires that one monitor the sign of the trial function as the walker is moved and reject steps that cause Ψ to change sign. In place of Algorithm 3.2, which was diffusion QMC with importance sampling, we have the following:

ALGORITHM 4.1 Fixed-node diffusion Monte Carlo.

```
C
C Choose an initial ensemble of size Nwalkers (see Algorithm 3.2). Before the
C walk begins, compute the total energy and trial function. As each electron is
C moved, the energy and trial function is updated.
C
      Do Istep=1,Nstep
         M = Nwalkers
         Do k=1,M
            Accept = 0.
            Do i=1,N
               CALL TRIALF(X,i,k,PSITX,ELOCALX,FQX)
C
C Move to a trial position and evaluate new quantities.
C
               Do j=1,3
                  Y(j) = X(i,j,k) + D*Tau*FQX(j) + GRAN(iseed)
               End Do
               CALL NEWTRIALF(Y,i,k,PSITY,ELOCALY,FQY)
C
C Check relative sign of PSIT
C
               If (PSITY / PSITX < 0 ) Then
                  A = 0
               Else
         ...Compute Metropolis acceptance probability as before...
               If (A > URAN(iseed)) Then
                  CALL COPY(Y, X(i))
                  Accept = Accept + 1/N
               End If
            End Do
C
C Compute branching and create or delete as before
C
                            .
                            .
                            .
            (see Algorithm 3.2)
                            .
                            .
                            .
```

```
End Do
End
```

One deficiency of the above algorithm is that an electron may take a path such as that shown in Fig. 4.2, which crosses and recrosses a node. Since this path is accepted for a large time step, but would not be allowed with a smaller time step, this is an artifact of our method and represents an additional time step bias known as the *cross-recross* error.

In Algorithm 4.1 we moved electrons one at a time, each time checking for node crossings. For the exact Green's function methods, all the electrons are accepted or rejected together, hence the cross-recross error is expected to be even larger than that associated with moving each electron separately. Therefore the fixed-node approximation introduces an additional time-step bias into these methods. (The exception is the domain GFMC method, in which each walker stays strictly within a domain, and can readily avoid the nodes.) In principle, we can correct for cross-recross by estimating the distance to the nearest node, using for example Newton's method, and restricting the step size as it approaches a node. In practice, one often employs the released node algorithm, discussed below, to eliminate the nodal bias altogether.

The fixed-node approximation has been widely applied. It has been shown to be highly accurate for small molecules with Ψ's that are readily available from standard *ab initio* calculations. To illustrate the accuracy of the fixed-node method, in Table 4.1 we show some representative results for first-row atoms. For heavier atoms, the fixed-node energy becomes less accurate due to the simplicity of the Ψ's used in these studies, which are typically no more than an SCF or Hartree-Fock determinant multiplied by an explicit electron-correlation function. Further trial function improvements should result in more accurate fixed-node results.

Figure 4.2: Example of cross-recross error. A large step takes a path that crosses and then recrosses the node. Such a path is rejected when using smaller time steps.

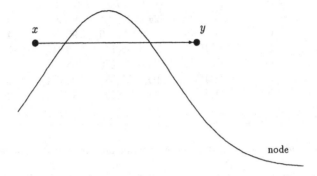

4.1.2 Understanding the nodes

Of central importance to the fixed-node approximation are the wave function nodes, which are typically less familiar and less well studied than the wave functions *per se*. Indeed, relatively little is known about the topology of the exact nodes. Moreover, even for a given trial function, the many-electron nodes of the wave function are difficult to visualize. The following discussion will attempt to clarify the subject somewhat.

For a many-electron system, the nodes of a wave function are the hypersurfaces on which the function vanishes. For a molecule or atom with N electrons, the nodes of Ψ will occupy a $3N - 1$ dimensional space, and will divide the $3N$ dimensional configuration space of the electrons into hypervolumes referred to as *nodal volumes* of Ψ. For an antisymmetric ground state wave function, every volume enclosed by anti-

Table 4.1: Total atomic energies computed by fixed-node QMC.

System	Method	QMC Energy	%CE	Ref.
Li	DMC	$-7.4784(1)$	100.0(5)	2
Be	DMC	$-14.66717(3)$	100.5(3)	3
B	Domain	$-24.6430(40)$	92(3)	4
C	Domain	$-37.8280(120)$	90(8)	4
N	DMC	$-54.5765(12)$	93.1(6)	5
O	Domain	$-75.0590(270)$	98(11)	4
F	DMC	$-99.7167(13)$	95.0(4)	6
Ne	DMC	$-128.9220(20)$	96.2(5)	3

symmetry nodes is equivalent. Specifically, the amplitude, gradient, and local energy of Ψ at two positions in space related by an exchange of two electrons are identical. Note that the antisymmetry constraint, Eq. 4.1, is not sufficient to determine the nodes; it defines only $3N - 3$ dimensional hyperpoints. This sparsity of information is unfortunate, because if Eq. 4.1 defined the nodes fully, any antisymmetric trial function would have the exact nodes.

The all-electron nodes do not correspond to the more familiar one-electron nodes of molecular orbitals. For example, a trial function for Li can be written as

$$\Psi = \det \begin{vmatrix} 1s(1) & 1s(2) \\ 2s(1) & 2s(2) \end{vmatrix} \alpha(1)\alpha(2) \det |1s(3)| \, \beta(3) \, . \tag{4.7}$$

where $1s(n)$ and $2s(n)$ represent respectively the 1s and 2s atomic orbitals occupied by electron n, and α and β are the spin functions. Here we have separated the electrons by spin, which is discussed further in Ch. 5. The nodes of Ψ, unlike the nodes of the orbitals, are the nodes of the determinant,

$$\det \begin{vmatrix} 1s(1) & 1s(2) \\ 2s(1) & 2s(2) \end{vmatrix} = 1s(1)2s(2) - 1s(2)2s(1). \tag{4.8}$$

Note that even though the 2s orbital has a radial node, the determinant is not zero

Figure 4.3: Nodes of linear H_3. The curves are sections through the nodal surfaces — the full nodal surfaces are obtained by revolving the curves about the internuclear axis. The trial function is zero whenever the like-spin electrons are both on any one of these surfaces.

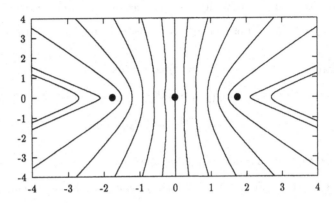

when the 2s orbital is zero. Rather, it is zero on the spherical shell $r_1 = r_2$, where r_n is the radial distance of electron n from the nucleus. Therefore, one must be cautious not to ascribe the nodes of an orbital to the trial function.

The nodes of three-electron doublet-coupled systems can be mapped in three dimensions. Such nodes are shown in Fig. 4.3 for H_3, derived from a single determinant trial function.[7] For many-electron systems, the higher dimensional surfaces are too complicated to be mapped in such a simple fashion.

4.1.3 Releasing the nodes

To proceed beyond the fixed-node approximation, we return to the problem that Φ_0^F, the Fermi ground-state wave function, cannot be immediately interpreted as a density. However, one can consider a general antisymmetric wave function Φ_A to be constructed from two non-negative symmetric functions Φ^+ and Φ^- such that

$$\Phi_A = \Phi^+ - \Phi^-. \tag{4.9}$$

Since both Φ^+ and Φ^- are positive, each separately can be interpreted as a density. If the Monte Carlo solution, $\phi_A(\mathbf{y}, \tau)$, is to be iterated in the usual manner,

$$
\begin{aligned}
\phi_A(\mathbf{y}, \tau + \delta\tau) &= \int G(\mathbf{y}, \mathbf{x}; \delta\tau)\phi_A(\mathbf{x}, \tau)\, d\mathbf{x} \\
&= \int G(\mathbf{y}, \mathbf{x}; \delta\tau)\phi^+(\mathbf{x}, \tau)\, d\mathbf{x} - \int G(\mathbf{y}, \mathbf{x}; \delta\tau)\phi^-(\mathbf{x}, \tau)\, d\mathbf{x} \\
&= \phi^+(\mathbf{y}, \tau + \delta\tau) - \phi^-(\mathbf{y}, \tau + \delta\tau),
\end{aligned}
\tag{4.10}
$$

we see that we can propagate ϕ_A by propagating two functions, ϕ^+ and ϕ^-, individually. It is easy to verify that both ϕ^+ and ϕ^- must have finite overlap with both the Fermi ground state and the lower energy Bose states. A simple choice of ϕ^+ and ϕ^- at the beginning of the walk is

$$\phi^\pm(t = 0) = \frac{1}{2}(|\Psi| \pm \Psi). \tag{4.11}$$

That is, $\phi^+(\tau = 0)$ is non-zero only in the positive regions of Ψ, and $\phi^-(t = 0)$ is non-zero only in the negative regions of Ψ. At large τ, after all the excited states have decayed, we are left with

$$\phi^\pm(\tau \to \infty) = \pm C_F \Phi_0^F + C_B \Phi_0^B e^{(E_0^F - E_0^B)\tau}. \tag{4.12}$$

Clearly $\phi^+ - \phi^- \propto \Phi_0^F$. But note that $E_0^F - E_0^B$, the difference between the Fermi and Bose ground state energies, always will be positive, causing the Fermi component asymptotically to become *exponentially* small as both ϕ^+ and ϕ^- converge separately to the Bose ground state (see Fig. 4.4).

Figure 4.4: Imaginary-time behavior of ϕ^+ and ϕ^-. At large τ small fluctuations in these functions can easily mask the "signal" $\phi^+ - \phi^-$.

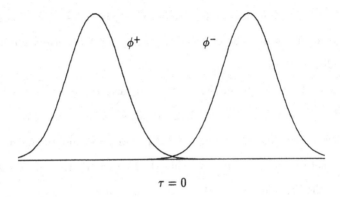

$\tau = 0$

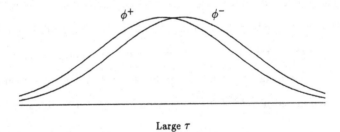

Large τ

The Bose component can be projected out of the energy, provided that Ψ is an antisymmetric trial function, viz

$$
\begin{aligned}
E_{\text{QMC}} &= \frac{\int \Phi_A(\tau \to \infty)\mathcal{H}\Psi d\mathbf{x}}{\int \Phi_A(\tau \to \infty)\Psi d\mathbf{x}} \\
&= \frac{\int \Phi^+ \mathcal{H}\Psi d\mathbf{x}}{\int(\Phi^+ - \Phi^-)\Psi d\mathbf{x}} - \frac{\int \Phi^- \mathcal{H}\Psi d\mathbf{x}}{\int(\Phi^+ - \Phi^-)\Psi d\mathbf{x}} \\
&= E_0^F ,
\end{aligned}
\tag{4.13}
$$

where Eq. 4.12 has been used. The second equality is written to emphasize that the Monte Carlo energy must be obtained as the energy difference of the positive and negative distributions.

The difficulty with this method is that its variance, and thus the noise, grows exponentially! From Eqs. 4.13 and 4.12 one can show the cause of this. The energy variance, which contains the average of $(\mathcal{H}\Phi_A)^2$, grows because the square of the Bose term does not cancel in the variance, while the Bose contribution grows exponentially relative to the Fermi term. For this reason this approach is classified as a *transient estimator* method. If the QMC energy decays to the Fermi ground state sufficiently quickly, one may obtain convergence prior to the Boson "noise" overwhelming the Fermion "signal." Whether this condition can be satisfied depends on the decay time of the excited states relative to the divergence of the exponential in Eq. 4.12. The decay of the excited states depends primarily on the energy gap between the ground and first-excited Fermi state of the same symmetry $E_0^F - E_1^F$. In particular, if

$$
E_1^F - E_0^F \gg E_0^F - E_0^B ,
\tag{4.14}
$$

then the transient estimator method should be successful. Nevertheless, the overall population will increase exponentially with the rate constant $E_0^F - E_0^B$. As the simulation continues, the noise will thus continue to grow; however, in the situation of Eq. 4.14 one will converge to E_0^F before the signal is overwhelmed.

The *released node* method[8] is a direct application of the above approach. By using an initial fixed-node distribution, one can help to stabilize the population for as long as possible. Once created, the fixed-node distribution $\{\mathbf{X}^{\text{fn}}\}$ is propagated as usual, with the exception that walkers are allowed to cross the nodes of Ψ. When a walker crosses such a node, it is placed in a "released node" population $\{\mathbf{X}^{\text{rn}}\}$, together with a count of the number of steps the walker has taken since it crossed the node (designated as s_k for walker k). This count is used to bin walkers by "age" and hence by variance. Each walker also contributes to the energy (or other property) with a sign, retained as a weight, ω_k, depending on the number of node crossings ($+1$ for even, -1 for odd).

To properly sample the nodal regions of Ψ, where the true wave function is not necessarily zero, it is necessary to use a guiding function Ψ_{G} distinct from Ψ. The guiding function is chosen to be positive everywhere; thus walkers may sample the nodes of Ψ. Recall from Ch. 2 that when using distinct guiding and trial functions, one samples the local energy $\mathcal{H}\Psi/\Psi$ from the distribution Ψ_{G}^2. All quantities evaluated at $\mathbf{X}_k$ must therefore be weighted by

$$W(\mathbf{X}_k) \equiv \Psi(\mathbf{X}_k)/\Psi_{\text{G}}(\mathbf{X}_k). \tag{4.15}$$

Therefore the fixed-node energy can be obtained as

$$E_0^{\text{fn}} = \frac{\sum_k W(\mathbf{X}_k^{\text{fn}}) E_L(\mathbf{X}_k^{\text{fn}})}{\sum_k W(\mathbf{X}_k^{\text{fn}})}, \tag{4.16}$$

while the released-node energy is the sum over all walks (whether they have crossed a node or not) with additional weight ω_k,

$$E^{\text{rn}} = \frac{\sum_k \omega_k W_k(\mathbf{X}_k^{\text{rn}}) E_L(\mathbf{X}_k^{\text{rn}})}{\sum_k \omega_k W_k(\mathbf{X}_k^{\text{rn}})}. \tag{4.17}$$

The estimation of the energy from Eq. 4.17 becomes difficult due to the rising variance as the "age" s_k increases. To avoid mixing walkers corresponding to different variances, one can form the more useful average

$$E^{\mathrm{rn}}(s_n) = \frac{\sum_{k:s \leq s_n} \omega_k W_k(\mathbf{X}_k^{\mathrm{rn}}) E_L(\mathbf{X}_k^{\mathrm{rn}})}{\sum_{k:s \leq s_n} \omega_k W_k(\mathbf{X}_k^{\mathrm{rn}})} \ , \tag{4.18}$$

where only those walks for which s_k is less than some value s_n are included in the sum. By plotting the sequence of energies $E^{\mathrm{rn}}(0)$, $E^{\mathrm{rn}}(1)$, $\cdots$, $E^{\mathrm{rn}}(s_{\mathrm{max}})$, one can observe convergence versus error growth. All walks are terminated when $s_k = s_{\mathrm{max}}$ in order to maintain a workable ensemble size (which is increasing due to the exponential in Eq. 4.12). This method has been successfully applied to a number of small atoms and molecules; some of these results are shown in Table 4.2.

4.1.4   Adaptive nodes

An alternative route for obtaining the exact nodes is to construct an appropriate stochastic procedure with this goal. One method[9] involves the construction of a separate *nodal function* which is used to separate walkers into positive and negative regions. The fixed node method uses an approximate analytic trial function for this purpose. Rather than using a trial function, we can construct the nodal function directly from the ensemble. For example, the nodal function can be constructed from an antisymmetrized sum of Gaussian functions centered on each walker. For an ensemble of M walkers with positions $\{\mathbf{X}\}$, such a nodal function, Ψ_{node}, is given by

$$\Psi_{node}(\mathbf{x}; \{\mathbf{X}\}) \equiv \mathcal{A}\left(\sum_{i=1}^{M} e^{-\|\mathbf{x}-\mathbf{X}_i\|^2}\right) \ , \tag{4.19}$$

where $\mathcal{A}$ is the antisymmetrizer operator and $\|\cdots\|$ gives the distance in the $3N$ dimensional electronic configuration space. The initial ensemble, $\{\mathbf{X}^{(0)}\}$, is chosen arbitrarily, perhaps from a preceding fixed-node walk. Each walker is then moved in the standard way, with the constraint that it not cross the nodes of $\Psi_{node}(\mathbf{x}; \{\mathbf{X}^{(0)}\})$. This move produces a new ensemble $\{\mathbf{X}^{(1)}\}$. The process is continued, updating Ψ_{node}

Table 4.2: Total energies of atoms and molecules computed by the Coulomb 1 GFMC method with released nodes.[8]

System	QMC Energy[a]	%CE
LiH	−8.071(1)	100(1)
Li$_2$	−14.994(2)	100(2)
H$_2$O	−76.43(2)	100(5)

(a) Energies in hartrees.

with each new ensemble. The result is an adaptive description of the nodes, that does not depend upon *a priori* knowledge of the wave function.

The main difficulties are the scaling of cost with N and M, and the accuracy of the method. Concerning scaling, there are $N!$ permutations and a large number, M, of walkers. Thus, computation of Ψ_{node} for any but the smallest systems is a huge effort. In addition, there is no theoretical reason to presume that the nodes have the structure of Eq. 4.19, and therefore that Ψ_{node} is capable of representing the exact nodes. Nevertheless, this method has been applied successfully to the first P state of H and the first triplet state of H$_2$.

4.2 Interacting Walker Methods

Given the difficulty in quantifying the nodes, and the transient nature of the released node method, other methods to obtain the exact energy are needed. In the methods described in the following, no information about antisymmetry is drawn from the nodes. Rather, the antisymmetry is built from positive and negative populations as was done with the released node method. The distinction between the released node method and the interacting ensemble methods may not be clear at first, especially as many of the latter methods begin with the former. The common feature of the interacting ensemble methods is the stabilization of the populations via imposed walker-walker interactions.

4.2.1 Pairing methods

One method for stabilizing the positive and negative populations involves *annihilation* of positive-negative walker pairs. For an infinite ensemble, the annihilation of positive and negative walkers occupying essentially the same position in $3N$-space would result asymptotically in exactly orthogonal $\phi^{\pm}$ functions. Annihilation both keeps the positive and negative distributions separated and orthogonal to the Bose component. This serves to stabilize the population growth and prevent the behavior depicted earlier in Fig. 4.4. With finite ensembles, however, two walkers will never occupy the same point. Rather, in the simplest form of annihilation one merely removes walker pairs occupying the same pre-defined volume element. For example, we can divide all of $3N$-dimensional space into hypercubes, and annihilate pairs of positive and negative walkers within each cube. This approach is impractical for more than a very few electrons, because the number of hypercubes required grows exponentially, while the density of walkers decreases exponentially with the dimensionality of space.

A more practical procedure is to compute the distance between a pair of walkers, and then to annihilate that pair based upon some probability which depends on this distance. A natural choice for this probability is the Green's function. Because the Green's function is the probability that a walker at $\mathbf{x}$ will move to $\mathbf{y}$, if the walker at $\mathbf{x}$ is positive, and the walker at $\mathbf{y}$ is negative, then $G(\mathbf{y}, \mathbf{x})$ is the probability that the positive walker will move to the position of the negative one, and hence annihilate.

One method to accomplish this annihilation is through the pair Green's function,[10]

$$G_p(\mathbf{y}, \{\mathbf{x}^+, \mathbf{x}^-\}) = G(\mathbf{y}, \mathbf{x}^+) - G(\mathbf{y}, \mathbf{x}^-), \tag{4.20}$$

where G_p is interpreted as the probability of the pair of walkers $(\mathbf{x}^+, \mathbf{x}^-)$ producing a walker at point $\mathbf{y}$. If G_p is positive, it is the probability that the walker at $\mathbf{y}$ is placed in the positive population; if G_p is negative, the resulting walker would be put into the negative population. (The initial positive and negative populations may be defined, e.g., as in Eq. 4.11.) If $G_p = 0$, the pair $(\mathbf{x}^+, \mathbf{x}^-)$ annihilates. However, in practice two walkers have a negligible probability of overlapping precisely; thus the effect of G_p is to produce a "partial cancellation" at each step. The greatest cancellation will occur when pairs are close together in configuration space.

A related but more general procedure[11] is to assign each walker a weight, as is done in pure DMC (Sec. 3.2.4). One then moves both walkers ($\mathbf{x}^+$ and $\mathbf{x}^-$) in the usual fashion. Let the weight at step i be W_i^+ for the positive walker and W_i^- for the negative walker. These weights are then updated by

$$W_{i+1}^{\pm} = \frac{\min(\pm W_i^+ G(\mathbf{y}^+, \mathbf{x}^+) \mp W_i^- G(\mathbf{y}^-, \mathbf{x}^-), 0)}{G(\mathbf{y}^{\pm}, \mathbf{x}^{\pm})}. \tag{4.21}$$

Walkers may be deleted or split based upon these weights rather than the QMC branching factor.

The pairs should be chosen to have maximum cancellation. But how does one choose which walkers to pair? In general, this decision requires computing all interparticle distances. However, assuming that the walkers do not move too far in a single step, a given pairing may be good for many steps (as determined empirically). In addition, one can use both the molecular point group and exchange symmetry to aid in pairing. For instance, if a molecule has a center of inversion or rotation axis, one can apply the appropriate symmetry operations to each walker to minimize the inter-walker distance. More generally, one can perform any even permutation of electrons to minimize this distance. Another method to maximize cancellation is called *self-cancellation*. An odd permutation (exchange of two electrons) changes the sign of a walker. The permuted walker then has the possibility to annihilate its unpermuted self. Self-cancellations are particularly simple to compute since the

Table 4.3: Interaction energies ΔE of collinear symmetric H_3 obtained from Bessel GFMC with annihilation.[12]

R_{H-H} (bohr)	ΔE (kcal/mole)[a]
1.600	12.59(7)
1.650	10.96(3)
1.700	9.94(2)
1.725	9.71(2)
1.757	9.61(1)
1.775	9.67(2)
1.800	9.77(2)
1.850	10.38(4)
1.900	11.38(5)

(a) $\Delta E = E_{H_3} - E_{H_2} - E_H$, where E_{H_2}
is the exact Born-Oppenheimer energy of H_2
and $E_H = 0.5$ hartrees.

inter-walker distance in $3N$ space is just the square-root of twice the interelectronic distance.

These methods show great promise, and have been applied to several small systems. In Table 4.3 and Fig. 4.5 we show computed interaction energies for the H + H_2 exchange reaction obtained using the Bessel method (cf. Ch. 3) with annihilation. These results provide the most accurate barrier height to date for this reaction. The main obstacle to the application of this method to large systems is the computational cost associated with efficient annihilation, i.e., the determination of the interwalker distances for a large number of walker pairs.

4.2.2 Fully interacting ensembles

The above pairing methods raise questions as to the optimal pairing strategy. A more consistent treatment would allow each walker to interact with the entire ensemble, rather than pairing specific positive and negative walkers.[13] To have all the walkers

Figure 4.5: Plot of interaction energies ΔE (in kcal/mole) of collinear symmetric H_3 obtained from Bessel GFMC with annihilation. Values are from Table 4.3.

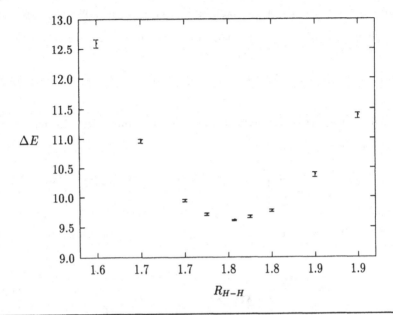

interact one can write

$$G_t(\mathbf{y}; \{\mathbf{X}\}) = \sum_{k=1}^{M} \{G(\mathbf{y}, \mathbf{X}_k^+) - G(\mathbf{y}, \mathbf{X}_k^-)\}. \qquad (4.22)$$

For the sum over the ensemble to be meaningful, there must always be an equal number of positive and negative walkers. This can be accomplished by either random deletion, or by randomly permuting positive to negative walkers. In analogy with Eq. 4.20, G_t acts as the probability of finding a new walker at $\mathbf{y}$. As before, positive walkers are produced when $G_t > 0$, and negative walkers are produced when $G_t < 0$. Unlike the previous method however, we will also choose two trial functions, Ψ^+ and Ψ^-, to help the two populations avoid each other. Recall that in importance sampling the trial function created a pseudo-force that pushed walkers toward regions of large $|\Psi|$. Hence we can do something similar here by choosing Ψ^+ to have strong overlap with the positive part of Φ and near zero overlap with the negative region, and similarly for Ψ^-. This causes the positive and negative walker populations to effectively repel each other. Then, at large time, the Monte Carlo solution is

$$\Phi = \Psi^+ \phi^+(t \to \infty) - \Psi^- \phi^-(t \to \infty), \qquad (4.23)$$

and is stable. For the Bessel method, the transition probability with importance sampling becomes

$$\tilde{G}_t(\mathbf{y}; \{\mathbf{X}\}) = \sum_{k=1}^{M} \left\{ \frac{G(\mathbf{y}, \mathbf{X}_k^+)}{\Psi^+(\mathbf{X}_k^+)} \frac{V(\mathbf{X}_k^+)}{E} - \frac{G(\mathbf{y}, \mathbf{X}_k^-)}{\Psi^-(\mathbf{X}_k^-)} \frac{V(\mathbf{X}_k^-)}{E} \right\} . \qquad (4.24)$$

Sampling from this distribution can be accomplished by rejection, based on the probability

$$p = \frac{\tilde{G}_t(\mathbf{y}; \{\mathbf{X}\})}{\sum\limits_{k=1}^{M} \left\{ \left| \frac{G(\mathbf{y}, \mathbf{X}_k^+)}{\Psi^+(\mathbf{X}_k^+)} \frac{V(\mathbf{X}_k^+)}{E} \right| + \left| \frac{G(\mathbf{y}, \mathbf{X}_k^-)}{\Psi^-(\mathbf{X}_k^-)} \frac{V(\mathbf{X}_k^-)}{E} \right| \right\}} . \qquad (4.25)$$

This method results in the desired stability, but at the cost of the computation of $M(M-1)/2$ inter-walker distances. For a large ensemble, say $M = 1000$ walkers, this requires 449,500 distance computations in the $3N$-dimensional configuration space of the electrons. This cost problem is particularly difficult to overcome, because the ensemble size needs to be large to avoid a finite ensemble size bias, which is inherent in this method. One way to eliminate this bias is to compute the energy as a function of $1/M$, and extrapolate to $1/M \to 0$, but this still requires very large ensembles for reliable results. Nevertheless, this method has been applied to the Li atom, among others, and yielded a total energy of $-7.4776(6)$ hartrees, in agreement with the estimated exact energy of -7.4781 hartrees. Like the pairing methods, the fully interacting ensemble method holds great promise and needs to be studied further.

Exercises

1. A particle in a one-dimensional box whose walls are at $\pm L$ has the ground state wave function,

$$\Phi_0(x) = L^{-\frac{1}{2}} \cos\left(\frac{\pi x}{2L}\right)$$

and energy,

$$E_0 = \frac{\pi^2}{8L^2}.$$

Construct the first antisymmetric state for *two* particles in this one-dimensional box, i.e. $\Phi_0(x_1, x_2)$, and determine the node location. Use the DMC fixed-node algorithm without importance sampling to see how well it replicates the analytic solution. (Check for node crossings either directly or with the sign of $\Phi_0(x_1, x_2)$.) In this simulation, what is the effect of moving the node on the energy?

2. Do the simulation in Exercise 1 with an incorrect node, and then perform a released node walk. Determine the energy and variance as a function of s_k by use of Eq. 4.18.

3. Modify the simulation in Exercise 2 to annihilate positive and negative walkers co-located within areas defined by a two-dimensional grid. Does this modification stabilize the solution? What is the effect of the grid spacing on accuracy and on computation time?

Suggestions for Further Reading

1. M. H. Kalos, "Optimization and the Many-Fermion Problem," in *Monte Carlo Methods in Quantum Problems*, 19-31 (D. Reidel, Dordrecht, 1984).

2. K. E. Schmidt and M. H. Kalos, "Few- and Many-Fermion Problems," in *Monte Carlo Methods in Statistical Physics II*, 125-44 (Springer-Verlag, New York, 1984).

3. B. H. Wells, "Green's Function Monte Carlo Methods," in *Methods in Computational Chemistry* **1**, 311-50 (Plenum, New York, 1987).

References

1. J. B. Anderson, "Quantum Chemistry by Random Walk. H^2P, $H_3^+ D_{3h}^1 A_1$, $H_2^3\Sigma_u^+$, $H_4^1\Sigma_g^+$, Be ^1S," *Journal of Chemical Physics* **85**, 4121-27 (1976).

2. R. N. Barnett, P. J. Reynolds and W. A. Lester, Jr., "Monte Carlo Determination of the Oscillator Strength and Excited State Lifetime for the Li $2^2S \rightarrow 2^2P$ Transition," *International Journal of Quantum Chemistry* **42**, 837-47 (1992).

3. C. J. Umrigar, M. P. Nightingale and K. J. Runge, "A Diffusion Monte Carlo Algorithm with Very Small Time-Step Errors," *Journal of Chemical Physics* **99**, 2865-90 (1993).

4. R. P. Subramaniam, M. A. Lee, K. E. Schmidt and J. W. Moskowitz, "Quantum Simulation of the Electronic Structure of Diatomic Molecules," *Journal of Chemical Physics* **97**, 2600-9 (1988).

5. P. J. Reynolds, R. N. Barnett, B. L. Hammond and W. A. Lester, Jr., "Molecular Physics and Chemistry Applications of Quantum Monte Carlo," *Journal of Statistical Physics* **43**, 1017-26 (1986).

6. D. R. Garmer and J. B. Anderson, "Quantum Chemistry by Random Walk: Application to the Potential Energy Surface for $F + H_2 \rightarrow HF + H$," *Journal of Chemical Physics* **86**, 7237-39 (1987).

7. R. N. Barnett, P. J. Reynolds and W. A. Lester, Jr., "H + H₂ Reaction Barrier: A Fixed-Node Quantum Monte Carlo Study," *Journal of Chemical Physics* **82**, 2700-7 (1985).

8. D. Ceperley and B. J. Alder, "Quantum Monte Carlo for Molecules: Green's Function and Nodal Release," *Journal of Chemical Physics* **81**, 5833-44 (1984).

9. R. Bianchi, D. Bressanini, P. Cremaschi and G. Morosi, "Antisymmetry in the Quantum Monte Carlo Method with the A-Function Technique: H_2 $b^3\Sigma_u^+$, H_2 $c^3\Pi_u$, He 1 ^{3}S," *Journal of Chemical Physics* **98**, 7204-9 (1993).

10. D. M. Arnow, M. H. Kalos, M. A. Lee and K. E. Schmidt, "Green's Function Monte Carlo for Few Fermion Problems," *Journal of Chemical Physics* **77**, 5562-72 (1982).

11. J. B. Anderson, C. A. Traynor and B. M. Boghosian, "Quantum Chemistry by Random Walk: Exact Treatment of Many-Electron Systems," *Journal Chemical Physics* **95**, 7418-25 (1991).

12. D. L. Diedrich and J. B. Anderson, "An Accurate Quantum Monte Carlo Calculation of the Barrier Height for the Reaction $H + H_2 \rightarrow H_2 + H$," *Science* **258**, 786-88 (1992).

13. S. Zhang and M. H. Kalos, "Exact Monte Carlo for Few-Electron Systems," *Physical Review Letters* **67**, 3074-77 (1991).

Chapter 5

Variational Trial Functions

Important to both VMC and QMC is the choice of the trial function. In VMC, all averages are evaluated with respect to the trial function, and so it determines the ultimate accuracy. In fixed-node QMC, the accuracy depends on the quality of the nodes of the trial function. Separate from the fixed-node approximation, the trial function affects the variance that can be achieved for a given amount of computation.

Most variational methods rely on a double basis-set expansion in 1-electron primitive functions and in N-electron Slater determinants. A unique characteristic of Monte Carlo methods is their ability to use arbitrary wave function forms — including ones with explicit interelectronic distance dependencies — enabling treatments beyond forms constructed solely with one-electron functions. Given this flexibility, it is important to recall properties a trial function ideally should possess. Thus, in this chapter we first review the known properties of exact solutions of the Schrödinger equation. We next discuss the various forms of approximate trial functions that currently are used. Much of this discussion draws on various independent-particle-based *ab initio* approaches that we consider as needed. In particular, we focus on electron correlation, and the capability of various variational forms of many-electron wave functions to describe it.

5.1   Properties of the Exact Wave Function

What do we know about the exact wave function? By definition, it satisfies $\mathcal{H}\Phi_k = E_k\Phi_k$. For bound electronic states, which are the focus here, we also know that in the absence of external fields Φ_k can be made real and is square integrable. In addition, Φ_k satisfies the virial, hypervirial, Hellmann-Feynman, and generalized Hellmann-Feynman theorems.[1] An important corollary of the eigenvalue equation is that the local energy is a constant for an eigenstate. For non-eigenstates, the variance of the local energy is an important measure of wave function quality.[2–4] It was used in Ch. 2 to help optimize the trial function. All of the above are global properties of Φ_k.

We also know several local properties of Φ_k, in addition to the above global properties. For example, because the local energy is a constant everywhere in space, each singularity of the Coulomb potential must be cancelled by a corresponding term in the local kinetic energy. This condition results in a *cusp*, i.e., a discontinuity in the first derivative of Φ_k where two charged particles meet. With a sufficiently flexible trial function one can include appropriate degrees of freedom which are then determined by the *cusp condition*.[5]

To examine the cusp condition, let us take the simplest case: the electron-nucleus cusp of a hydrogenic atom. This is shown in Fig. 5.1. For hydrogenic atoms one can write the exact wave function in spherical polar coordinates, (r, θ, ϕ), as the product of a radial function, $R(r)$, and an angular function, $\Omega(\theta, \phi)$ (which is related to the spherical harmonics), i.e.,

$$\Phi_k(r, \theta, \phi) = R_k(r)\,\Omega_k(\theta, \phi). \tag{5.1}$$

In focusing on the electron-nucleus cusp, we are only interested in the behavior of Φ as $r \to 0$. So let us examine the radial function, which is the solution of the radial

Schrödinger equation,

$$\left(\frac{d^2}{dr^2} + \frac{2}{r}\frac{d}{dr} + \frac{2Z}{r} - \frac{l(l+1)}{r^2} + 2E\right) R(r) = 0 , \tag{5.2}$$

where Z is the atomic charge. For $l = 0$ (i.e., S states) we see that the two r^{-1} terms must cancel, which leads to the cusp condition

$$\left(\frac{1}{R(r)}\frac{dR(r)}{dr}\right)\bigg|_{r=0} = -Z. \tag{5.3}$$

If we solve Eq. 5.3 for R, we find that R must be an exponential in r (see Fig. 5.1). What of the $l \neq 0$ case, and what happens to Eq. 5.3 if $R(0) = 0$? Let us treat the latter issue first and show that it addresses the former.

To treat the case of $R \to 0$ at the origin, we can explicitly factor out the leading r dependence, i.e.,

$$R(r) = r^m \rho(r) , \tag{5.4}$$

so that $\rho(r)$ is a function that does not go to zero at the origin. The kinetic energy terms in Eq. 5.2 become

$$r^m \rho'' + \frac{2(m+1)}{r}r^m \rho' + \frac{m(m+1)}{r^2}r^m \rho , \tag{5.5}$$

where we have abbreviated the first and second derivatives of ρ as ρ' and ρ'', respectively. Substituting these terms into Eq. 5.2, we find that the r^{-2} terms cancel if $m = l$, yielding the following equation for ρ,

$$\left(\frac{2(l+1)}{r}\frac{\rho'}{\rho} + \frac{2Z}{r} + \frac{\rho''}{\rho} + 2E\right) r^l \rho = 0. \tag{5.6}$$

The choice of $m = l$ should not be altogether too surprising, for it corresponds to the known behavior of the eigenfunctions of the hydrogenic atoms with respect to angular momentum. Now, by equating the r^{-1} terms in Eq. 5.6 one gets the general electron-nucleus cusp condition,

$$\frac{\rho'}{\rho}\bigg|_{r=0} = -\frac{Z}{l+1}. \tag{5.7}$$

Figure 5.1: The electron-nucleus cusp in a hydrogenic atom.

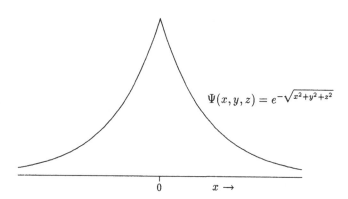

$$\Psi(x,y,z) = e^{-\sqrt{x^2+y^2+z^2}}$$

For hydrogenic systems, this cusp condition uniquely determines the overall exponential behavior of the wave function for each value of l, i.e, $R(l = 0) \propto e^{-Zr}$, $R(l = 1) \propto e^{-Zr/2}$, $R(l = 2) \propto e^{-Zr/3}$, etc.

The extension to the many-electron case is straightforward. As any single electron approaches the nucleus (with all others fixed), the exact wave function behaves asymptotically as in the one-electron case, and Eq. 5.7 holds for each electron individually. In the same way, an extension of this argument to the electron-electron cusp is possible. In this case, as electron i approaches electron j one has essentially a two-body problem similar to the hydrogenic atom. However, now both electrons give kinetic energy contributions. Expanding the wave function in spherical polar coordinates centered on electron i, leads to the "radial" equation,

$$\left(2\frac{d^2}{dr_{ij}} + \frac{4}{r_{ij}}\frac{d}{dr_{ij}} + \frac{2}{r_{ij}} - \frac{l(l+1)}{r_{ij}^2} + 2E\right) R_{ij}(r_{ij}) = 0. \tag{5.8}$$

Therefore, following the discussion for the hydrogenic systems, we obtain the cusp condition to be

$$\left.\frac{\rho'_{ij}}{\rho_{ij}}\right|_{r_{ij}=0} = \frac{1}{2(l+1)}. \tag{5.9}$$

We see that with a flexible enough form of Ψ, we can satisfy both Eqs. 5.7 and 5.9 thereby matching the Coulomb potential for any particle pair with terms from the kinetic energy.

Three-particle coalescence conditions also have been studied. These singularities are not a result of the potential, but are entirely due to the kinetic energy. The derivation of these terms is beyond our present scope. To provide a feel for the nature of these terms, we note that Fock (see the discussion in Ref. 6) showed by an examination of the helium atom in hyperspherical coordinates that terms of the form

$$(r_1^2 + r_2^2)\ln(r_1^2 + r_2^2) \tag{5.10}$$

are important when r_1 and $r_2 \to 0$ simultaneously. Additional higher-order terms, describing correlation effects and higher n-body coalescences, also have been suggested.

5.2 General Trial Function Forms

The exact wave function can be approximated in a number of ways through series expansions in the electronic coordinates. The convergence of these series depends upon the type of terms included. Hylleraas[7] and Pekeris[8] had great success for He with trial functions of the form

$$\Psi_{\text{Hylleraas}} = \left(\sum_{k=1}^{N} d_k r^{a_k} s^{b_k} t^{e_k}\right) e^{-\frac{1}{2}s}, \tag{5.11}$$

where r is the electron-electron separation (which we have previously designated as r_{ij}), $s = r_1 + r_2$, and $t = r_1 - r_2$. Here r_1 and r_2 are the scalar distances of the electrons from the nucleus. The electron-nucleus cusp condition is satisfied by the exponential term, and the electron-electron cusp is satisfied by choosing the proper

Table 5.1: Selected forms of many-electron wave functions with explicit interelectron distance dependence.

Form	Ψ
Hylleraas[7]	$e^{-\epsilon s}\sum_{\mu}c_{\mu}r^{l_{\mu}}s^{m_{\mu}}t^{n_{\mu}}$
Conroy[9, 10]	$\sum_{\mu}\frac{r_{12}^{\mu}}{n!(n+1)!}e^{b\sum_{i<j}(r_1^2+r_2^2+r_{12}^2)(r_1^2+r_2^2+2s)^{-1/2}}\sum_{mn}C_{mn}\varphi_m(1)\varphi_n(2)$
Ho[11]	$e^{\epsilon(r_1+r_2+r_3)}\sum_{\mu}c_{\mu}r_1^{l_{\mu}}r_2^{m_{\mu}}r_3^{n_{\mu}}r_{12}^{p_{\mu}}r_{13}^{q_{\mu}}r_{23}^{s_{\mu}}$
Frankowski & Pekeris[12]	$e^{-\epsilon s}\sum_{\mu}c_{\mu}r^{l_{\mu}}s^{m_{\mu}}t^{n_{\mu}}[(r_1^2+r_2^2)\ln(r_1^2+r_2^2)]$
Baker & Morgan[13]	$e^{-\epsilon s}\sum_{\mu}c_{\mu}r^{l_{\mu}}s^{m_{\mu}}t^{n_{\mu}}(\ln s)^{i_{\mu}}\sinh^{j_{\mu}}(b_{\mu}t)\cosh^{k_{\mu}}(b_{\mu}t)$

values for the coefficients. Because all the interparticle distances are represented, very accurate descriptions of the He wave function may be obtained with relatively few terms. Although Eq. 5.11 is written explicitly for two electrons, it is readily generalized to larger systems. More complicated forms of Ψ have been examined,[6-10] motivated primarily by the goal of satisfying the various n-body coalescences. These forms are summarized in Table 5.1.

Helium has long served as a testing ground for atomic trial functions because of its simplicity. One measure of the quality of a trial function form is the rate of convergence of the variational energy with the number of terms in the series. For example, a nine-term Hylleraas function yields an energy of –2.9035 hartrees, while a 1078 term function yields –2.903724375 hartrees.[8] Clearly, convergence is not fast. On the other hand, by adding terms with powers of $\ln s$ and negative powers of s, one can obtain –2.903724377033 hartrees with only 246 terms. The functional form clearly is very important. Recently, terms of the form $\cosh t$ and $\sinh t$ have been added[6] to model "in-out" correlation (the tendency for one electron to move away from the nucleus as the other approaches the nucleus). With 476 terms, an energy of –2.903724377034117 hartrees has been obtained. It was argued that comparable accuracy in a CI calculation would require 2×10^5 one-electron basis functions and approximately 7×10^{12} configurations.

Even though such accuracy is not typically sought, electron-correlation terms may provide clues for constructing Ψ's for many-electron systems. The difficulty with these general expansions for larger atoms and molecules is in constructing the correct spatial and exchange symmetries. Moreover, even with Monte Carlo integration, the task of determining hundreds or thousands of parameters remains an obstacle.

5.3   Hartree-Fock and Beyond

Unlike the forms just discussed, the most widely used methods in *ab initio* electronic structure theory are based on molecular orbital (MO) expansions and the Hartree-Fock approximation. MO theory has been the foundation for most theoretical chemical concepts during the past 50 years. As a first approximation, the N-electron wave function $\Phi(\mathbf{x}_1, \cdots, \mathbf{x}_N)$ is represented by a Slater determinant of spin orbitals. It often is abbreviated by writing only the diagonal elements of the Slater matrix, namely,

$$\Psi_D = \det \begin{vmatrix} \tilde{\varphi}_1(\mathbf{x}_1) & \cdots & \tilde{\varphi}_1(\mathbf{x}_n) \\ \vdots & \ddots & \vdots \\ \tilde{\varphi}_n(\mathbf{x}_1) & \cdots & \tilde{\varphi}_n(\mathbf{x}_n) \end{vmatrix} \equiv \det |\tilde{\varphi}_1(\mathbf{x}_1) \cdots \tilde{\varphi}_n(\mathbf{x}_n)| . \qquad (5.12)$$

Each spin orbital, $\tilde{\varphi}_i$, consists of a spatial function, φ_i, multiplied by an electron spin function (α or β). The orbital approach is motivated by a simple generalization of the one-electron description of the H atom, building in the antisymmetry required by the Pauli principle. The determinantal part of the probability distribution, Ψ_D^2, depends on the product $\varphi_1^2 \varphi_2^2 \cdots \varphi_n^2$ of one-electron probabilities. Since no terms involve conditional probabilities of two or more electrons, each particle acts independently of the others in this type of wave function, and the total probability is a simple product of one particle probabilities. This *independent-particle approximation* differs fundamentally from Hylleraas-type functions which include r_{ij} terms explicitly. The consequences of this approximation will be discussed next.

5.3.1   Hartree-Fock and correlation energies

In the independent-particle approximation, the expectation value of the energy can be broken into terms that are sums of integrals over the MOs. For simplicity, in the following discussion we assume a closed-shell system consisting of $2n$ electrons. Then the energy is given by the expression

$$E = 2 \sum_{i=1}^{n} H_{ii} + \sum_{i=1}^{n} \sum_{j=1}^{n} (2J_{ij} - K_{ij}), \tag{5.13}$$

where the integrals H_{ii}, J_{ij}, and K_{ij} are the matrix elements of the one-electron, Coulomb, and exchange integrals,[14] namely,

$$H_{ii} = \int \varphi_i(\mathbf{x}_1) \mathcal{H}^{\text{core}} \varphi_i(\mathbf{x}_1) d\mathbf{x}_1 \tag{5.14}$$

$$J_{ij} = \int \varphi_i(\mathbf{x}_1) \varphi_j(\mathbf{x}_2) \frac{1}{r_{ij}} \varphi_i(\mathbf{x}_1) \varphi_j(\mathbf{x}_2) d\mathbf{x}_1 d\mathbf{x}_2 \tag{5.15}$$

$$K_{ij} = \int \varphi_i(\mathbf{x}_1) \varphi_j(\mathbf{x}_2) \frac{1}{r_{ij}} \varphi_j(\mathbf{x}_1) \varphi_i(\mathbf{x}_2) d\mathbf{x}_1 d\mathbf{x}_2. \tag{5.16}$$

The one-electron Hamiltonian, $\mathcal{H}^{\text{core}}$, is the Hamiltonian without the two-electron potential terms. The application of the variational method to Eq. 5.13 yields the Hartree-Fock equations for the MOs and the orbital energies ϵ_i. Specifically, one obtains in place of the Schrödinger equation the following eigenvalue equation:

$$\mathcal{F} \varphi_i = \epsilon_i \varphi_i, \tag{5.17}$$

where $\mathcal{F}$ is the Fock operator

$$\mathcal{F} = \mathcal{H}_1 + \sum_{j=1}^{n}(2\mathcal{J}_j - \mathcal{K}_j), \qquad (5.18)$$

and the operators $\mathcal{H}_1$, $\mathcal{J}$, and $\mathcal{K}$ are defined such that $\mathcal{H}_1\varphi_i = H_{ii}$, $\mathcal{J}_j\varphi_i = J_{ij}$, and $\mathcal{K}_j\varphi_i = K_{ij}$. Through the action of the $\mathcal{J}$ and $\mathcal{K}$ operators, each MO is coupled to the mean field generated by the rest of the MOs. This mean field describes the behavior of an electron in the average field of the remaining electrons. In the Hartree-Fock approach, then, one has lost all information concerning pair- and higher-electron correlation. In accordance with the variational principle, the effect of neglecting these terms is to raise the energy above the exact energy.

The *correlation energy* is defined as the magnitude of the energy difference between the exact (non-relativistic) energy and the Hartree-Fock energy. To the extent that correlation energy is important, the success of the Hartree-Fock approximation arises often as a consequence of the cancellation of correlation energies between two systems of interest, e.g., between the energies of a neutral system and its positive ion in the case of an ionization potential. Unfortunately, one often is unable to predict *a priori* whether such a cancellation of correlation energies will occur. Therefore, for predictive calculations one typically must go beyond the Hartree-Fock method, to an approach that provides an estimate of the correlation energy.

5.3.2 Linear combination of atomic orbitals and the self-consistent field equations

Before leaving the Hartree-Fock method, let us complete the discussion of some computational details, because so many methods, including Monte Carlo with trial functions, rely on the MOs produced by this method. Although Eqs. 5.13 and 5.17 give expressions for the (closed shell) energy and corresponding MOs, for many applications one desires an explicit representation of the MOs. For atoms and diatomic molecules, one can solve the Hartree-Fock equations on a grid, which is the essence of the numerical Hartree-Fock approach. While numerical Hartree-Fock yields exact Hartree-Fock MOs, it has yet to be achieved for arbitrary polyatomic molecules. A more generally applicable approach is to express the MOs as linear combinations of atomic orbitals (AOs). Specifically, each MO, φ_i, is expanded in an atomic orbital *basis set* of known functions, χ_μ,

$$\varphi_i(\mathbf{x}) = \sum_\mu c_{i\mu} \chi_\mu(\mathbf{x}). \tag{5.19}$$

(In the following, i and j refer to MOs and μ, ν, λ and σ refer to AOs.) The problem now has been changed into that of finding the MO coefficients, $c_{i\mu}$. Expressed in matrix form, the Hartree-Fock equations in an AO basis becomes the self-consistent field (SCF) equations

$$\mathbf{FC} = \mathbf{SCE} , \tag{5.20}$$

where $\mathbf{E}$ is a diagonal matrix of the orbital energies, $\mathbf{S}$ is the AO overlap matrix ($S_{\mu\nu} = \langle \chi_\mu | \chi_\nu \rangle$), $\mathbf{C}$ is the matrix of AO coefficients, $c_{i\mu}$, and $\mathbf{F}$ is the Fock matrix in the AO space, with matrix elements

$$F_{\mu\nu} = H_{\mu\nu} + \sum_{\lambda\sigma} \left(\sum_i c_{i\lambda} c_{i\sigma} \right) \left[(\mu\nu|\lambda\sigma) - \frac{1}{2}(\mu\lambda|\nu\sigma) \right]. \tag{5.21}$$

The quantities $(\mu\nu|\lambda\sigma)$ are shorthand notation for the J and K two-electron integrals in terms of AOs, namely,

$$(\mu\nu|\lambda\sigma) = \int \chi_\mu(1)\chi_\nu(1)\frac{1}{r_{12}}\chi_\lambda(2)\chi_\sigma(2)d\mathbf{x}_1 d\mathbf{x}_2. \tag{5.22}$$

The most popular forms for the atomic functions are Slater-type orbitals (STOs), $Y_{lm}(\Omega)r^n e^{-\zeta r}$, where Y_{lm} are spherical harmonics and ζ is a variational parameter, and Cartesian Gaussian-type orbitals (GTOs), $x^i y^j z^k e^{-\zeta r^2}$. The STOs are chosen because they resemble the hydrogenic atomic orbitals, whereas the GTOs are most widely used for molecules because they are much simpler to integrate analytically. These basis sets are defined by the set of functions used, and the values of l, m and n, or i, j and k, and the exponents ζ for each one.

5.3.3   Cusp conditions

In Monte Carlo, we may factor the determinant into electron spin components, i.e., $\Psi_D = \Psi_\alpha \Psi_\beta$. This factorization is justified because all possible permutations of this term that are present in the full antisymmetric Ψ are operationally equivalent: each term simply corresponds to a relabeling of the electrons. Expectation values, and even local values of any observable, are unchanged.

Given this factorization, let us now determine the cusp of the Hartree-Fock determinant. Since Ψ_D does not depend upon r_{ij} when i and j are electrons of opposite spin, the derivative with respect to r_{ij} must be zero and Ψ_D cannot satisfy the cusp condition. For parallel-spin electrons, however, even though the orbitals and Ψ_D do not depend explicitly upon r_{ij}, the determinant goes to zero as two electrons approach each other. The determinant columns containing $\mathbf{x}_i$ and $\mathbf{x}_j$ can be re-expressed as functions of $\mathbf{x}_i$ and the spherical polar coordinate $\mathbf{r}_{ij}$. If we expand the determinant about $\mathbf{x}_j = \mathbf{x}_i$, or equivalently around $r_{ij} = 0$, and ignore angular and radial terms beyond the linear, we obtain

$$
\begin{aligned}
\Psi_D(\mathbf{x}_i, \mathbf{x}_j, \ldots) &= \Psi_D(\mathbf{x}_i, \mathbf{x}_i + \mathbf{r}_{ij}, \ldots) \\
&= \Psi_D(\mathbf{x}_i, \mathbf{x}_j = \mathbf{x}_i, \cdots) + r_{ij} \frac{\partial \Psi_D}{\partial r_{ij}}\bigg|_{r_{ij}=0} + \ldots
\end{aligned}
\tag{5.23}
$$

The term $\Psi_D(\mathbf{x}_i, \mathbf{x}_j = \mathbf{x}_i, \cdots)$ is zero because two electrons have identical coordinates. Except for special cases, the derivative $\partial \Psi_D / \partial r_{ij}|_{r_{ij}=0}$ is non-zero. We conclude that for small r_{ij} we may write the determinant in the form $\Psi_D = r_{ij}\rho_{ij}$, similarly to what was earlier done in Sec. 5.1. Following that notation, we have here that $l = 1$.

However, since the factor ρ_{ij} has no dependence upon r_{ij}, we cannot satisfy the cusp condition, Eq. 5.9. Hence a single determinant wave function alone does not satisfy the electron-electron cusp condition for either unlike or like spins. For later use, we note that Ψ_D treats like- and unlike-spins differently, requiring $l = 1$ and $l = 0$, respectively.

In contrast, Ψ_D *can* be made to satisfy the electron-nucleus cusp condition because the AOs explicitly depend upon r_{iA}. Consider the most common case, where Ψ_D does not go to zero at $r_{iA} \to 0$. Then,

$$\frac{1}{\Psi_D} \frac{\partial \Psi_D}{\partial r_{iA}}\bigg|_{r_{iA}=0} = -Z_A. \tag{5.24}$$

How can we enforce Eq. 5.24? First note that the derivative of a determinant is equal to the sum of determinants in which each column of the original determinant is sequentially substituted with its derivative. Since each column of the Slater matrix involves only the coordinates of a single electron, $\partial \Psi_D / \partial r_{iA}$ is equal to just one of these terms, namely the determinant in which column i is replaced by $\partial \varphi_k / \partial r_{iA}$. As Eq. 5.24 indicates, the cusp condition requires that the derivative of the determinant divided by the determinant be a constant, *regardless* of the positions of the other electrons. This can only be accomplished if for each MO (i.e., for each k) one requires

$$\frac{1}{\varphi_k} \frac{\partial \varphi_k}{\partial r_{iA}}\bigg|_{r_{iA}=0} = -Z_A. \tag{5.25}$$

The common multiplier Z_A factors out of the determinant, and the numerator and denominator cancel. Therefore, for Ψ_D to satisfy Eq. 5.24, each orbital must *individually* satisfy Eq. 5.25. When an orbital φ_k is zero at the origin, one may use the general expression Eq. 5.7. Note, however, that unless the entire determinant is zero at the origin, MOs that are zero there do not contribute to the total cusp behavior. This can be seen by multiplying both sides of Eq. 5.25 by φ_k. The right– and left-hand sides will vanish no matter what value is used for Z_A. Note also that while the STOs have the proper hydrogenic cusp behavior, the GTOs do not. Basis sets consisting of GTOs, although computationally expedient for integral evaluation, cannot satisfy the electron-nucleus cusp condition, and are therefore generally unsuitable for Monte Carlo trial functions.

5.3.4 Some Hartree-Fock trial function properties

Several features of the Hartree-Fock and SCF methods are worth summarizing here, since they are of interest for constructing Monte Carlo trial functions:

1. SCF computations will only reproduce Hartree-Fock results in the limit of an infinite AO basis set. Traditionally this leads to a number of difficulties associated with incompleteness of basis sets, such as further raising of the energy, basis set superposition error, and failure to satisfy the virial and Hellmann-Feynman theorems. These problems are less significant in QMC, but still may play a role.

2. Traditional SCF methods require diagonalization of an $N_{AO} \times N_{AO}$ matrix. Diagonalization requires on the order of N_{AO}^3 operations.

3. Solution of Eq. 5.20 is obtained by an iterative procedure. Since the MOs must be known in order to construct the $F_{\mu\nu}$ matrix, an initial guess must be made to $C_{i\mu}$. The process is iterated to self-consistency.

4. The two-electron integrals, Eq. 5.22, involve four AOs. Thus, their computation is an order N_{AO}^4 process. However, these integrals need to be computed only once (for a given state and nuclear geometry) and are readily computed in parallel.

5.3.5   Post Hartree-Fock methods

There are two broad categories of post-Hartree-Fock methods, i.e. methods that go beyond Hartree-Fock: configuration interaction (CI) and many-body perturbation theory (MBPT). Because our focus here is in obtaining variational trial functions for QMC, we shall not discuss perturbation theory, although wave functions constructed from perturbation theory could be used.

In CI one begins by noting that the *exact* N-electron wave function can be expanded as a linear combination of an infinite set of Slater determinants, $\{D_I\}$, i.e.

$$\Psi = \sum_{I=0}^{\infty} c_I D_I, \tag{5.26}$$

where the D_I span the Hilbert space of electrons. The D_I can be any complete set of N-electron antisymmetric functions. One such choice is obtained from the Hartree-Fock method by substituting all excited states for each MO. This, of course, requires an infinite number of determinants, derived from an infinite AO basis set, including continuum functions.

To understand CI better we need to consider correlation in more detail. There are essentially two kinds of correlation: geometrical and dynamical. First, let us consider dynamical correlation. It is manifested, for example, in the angular correlation in He. The Hartree-Fock determinant places both electrons uniformly in spherical symmetry around the nucleus, that is, the electrons are uncorrelated. To correct this, one can add a small contribution of a determinant of P symmetry to obtain,

$$\Psi = c_1 \det|1s(1)1s(2)| + c_2 \det|1s(1)2p(2)|. \tag{5.27}$$

If c_2 is properly chosen, the second term increases the wave function when the electrons are on opposite sides of the nucleus, and decreases Ψ when they are on the same side. Likewise, radial correlation can be achieved by adding a $2S$ He "configuration" into

Ψ. Both of these correlation terms describe (in part) the instantaneous positions taken by the two electrons, and contribute to the *dynamic* correlation. *Non-dynamic* or *geometrical* correlation, on the other hand, is encountered in the dissociation of a molecule — i.e., when the geometry of the molecule changes. It also occurs when a Hartree-Fock excited state is close enough in energy to mix with the ground state. For example, a well known deficiency of the Hartree-Fock determinant is that it does not dissociate into two neutral fragments, but rather into ionic configurations. For a proper description of reaction pathways, a multi-determinant wave function is required — one containing a determinant or a linear combination of determinants corresponding to all fragment states.

To illustrate the relative roles of dynamic and non-dynamic correlation let us consider the Be atom. Although a single determinant together with a very flexible Padé-Jastrow function (see Sec. 5.4) can recover 100% of the correlation energy for two-electron systems,[15] for Be only 76% of the correlation energy is recovered with a similar trial function. However, when the non-dynamical contribution of $2p$ orbitals is added to the wave function, 99% of the correlation energy can be obtained. This effect persists, though to a lesser extent, in fixed-node QMC calculations on Be, which obtain only 89% of the correlation with a single determinant, but 100% with the inclusion of the $2p$ states.[15]

Note that no r_{ij} terms are included in CI expansions. The two-electron cusp is satisfied only asymptotically. The convergence of CI expansions has been analyzed in the literature.[6] Ideally, in an expansion one wishes to observe exponential convergence, i.e., the contribution of each new term in the series should be smaller by roughly a constant factor. However for CI expansions this convergence has been estimated to go only as $(L+1)^{-3}$, where L is the number of terms.[6] This is extremely slow.

The basic CI idea has sparked numerous related schemes designed to be computationally more practical, while retaining as much of the *differential* correlation energy (the difference in correlation between two systems) as possible. We list here some of the more common approaches that may be used, in principle, to construct VMC and QMC trial functions.

Full CI : The wave function is a linear combination of determinants of all possible excitations of the electrons into the virtual orbital space of the SCF basis. The number of such excitations is determined by the size of the AO basis set. Such calculations grow quickly in size as the size of the basis set is increased. Only small molecules have been treated to date.

Singles and Doubles CI (SDCI): Only single- and double-excitations of the electrons into the virtual orbitals are allowed. This makes possible the use of larger AO basis sets, but neglects the effects of three-body and higher correlations.

Multiconfiguration SCF (MCSCF): This method differs from the above methods in that the SCF or Hartree-Fock orbitals are not used to form the determinant basis set. Instead, starting from an SCF solution, both the CI and AO coefficients are optimized to obtain the lowest energy. This process requires more computation than CI alone, but the MCSCF MOs are more accurate than the Hartree-Fock MOs.

Complete Active Space SCF (CASSCF): This is a special case of MCSCF. Here a subset of orbitals are labeled chemically *active* while the rest are frozen. Both occupied and virtual orbitals may be used. Then a full CI calculation is carried out in the active space, and the MO coefficients are optimized. The choice of active space is made on the basis of which electrons are the primary determiners of the chemical process to be described, based on physical insight. For example, one might have as active electrons those involved in bond formation or low-lying electronic transitions.

Multireference CI (MRCI): Rather than taking only Slater determinants generated from the Hartree-Fock ground state, as in ordinary CI, one selects a *set* of reference determinants from which excitations will be performed. This makes possible a better description of molecules that are poorly described in the Hartree-Fock approximation. Moreover, if the reference space contains, e.g., double excitations from the ground state, then an SDCI performed from this state will include triple and quadruple excitations in the N-electron expansion. Typically the reference states are obtained by MCSCF or CASSCF expansions.

5.4 Correlated Molecular Orbital Functions

In the above discussion we have learned the importance of r_{ij} terms. On the other hand Hartree-Fock and post Hartree-Fock wave functions, which do not explicitly contain these terms, lead to molecular integrals that are substantially more convenient for integration. At present, the vast majority of work is done with the latter independent-particle-type functions. Correlated molecular orbital (CMO) methods incorporate the best of both.

Although there are many ways one might construct such CMO functions, typically they are constructed as a determinant of orbitals where each orbital is multiplied by a function of the interelectronic coordinates $f(r_{ij}) \equiv f_{ij}$. Explicitly,

$$\Psi_{\mathrm{CMO}} = \mathcal{A}\varphi_1(\mathbf{x}_1)\varphi_2(\mathbf{x}_2)\cdots\varphi_n(\mathbf{x}_n)f_{12}f_{13}f_{23}\cdots f_{n-1,n} \ , \qquad (5.28)$$

where $\mathcal{A}$ is the antisymmetrizer operator. Most often the form of f_{ij} is independent of the orbitals, making Ψ_{CMO} a simple product function

$$\Psi_{\mathrm{CMO}} = \Psi_D \Psi_C, \qquad (5.29)$$

where Ψ_D is the determinantal part and Ψ_C is a product of correlation functions.

One advantage of this form is that we can now satisfy the electron-electron cusp condition, because f_{ij} depends explicitly on r_{ij}. Using the results of Sec. 5.3.3 describing the r_{ij} dependence of Ψ_D, we obtain the cusp of Ψ_{CMO} by constraining Ψ_C so that for like-spin electrons,

$$\frac{1}{\Psi_C}\frac{\partial \Psi_C}{\partial r_{ij}}\bigg|_{r_{ij}=0} = \frac{1}{4},\qquad (5.30)$$

while for opposite spin electrons,

$$\frac{1}{\Psi_C}\frac{\partial \Psi_C}{\partial r_{ij}}\bigg|_{r_{ij}=0} = \frac{1}{2}.\qquad (5.31)$$

We can distinguish between two classes of functions Ψ_C. In the first class, Ψ_C contains polynomials in r_{ij}, similar to the Hylleraas functions, such as in the Conroy form[9, 10] given in Table 5.1. The second class is an exponential or Jastrow form[16, 17]

$$\Psi_C = \epsilon^U \;,\qquad (5.32)$$

where U contains all the r_{ij} dependence. Representative functional forms are shown in Table 5.2. Note that each form contains one or more parameters which can be used to represent the two-electron cusp.

As an example, consider the most commonly used form in QMC, the Padé-Jastrow function,

$$U = \sum_{i=1}^{N}\sum_{j<i}^{N}\frac{a_1 r_{ij} + a_2 r_{ij}^2 + \cdots}{1 + b_1 r_{ij} + b_2 r_{ij}^2 + \cdots}\;.\qquad (5.33)$$

The general behavior of e^U is shown in Fig. 5.2, beginning at unity for $r_{ij} = 0$ and asymptotically approaching a constant value for large r_{ij}. In the simplest case, where only a_1 and b_1 are non-zero, this asymptotic value is $\exp(a_1/b_1)$. One can verify that the electron-electron cusp condition requires a_1 to be $1/2$ for unlike spins and $1/4$ for like spins. The linear Padé-Jastrow form has only one free parameter, namely b_1,

Table 5.2: Comparison of different forms of the Jastrow U function.

Form	U
Padé-Jastrow[15]	$\sum\limits_{i,j,A} \frac{P(r_{iA}, r_{jA}, r_{ij})}{1+Q(r_{iA}, r_{jA}, r_{ij})}$
Boys-Handy[18]	$\sum\limits_{i,j,A} \sum\limits_{\mu} c_\mu \left(\frac{a_{1\mu} r_{iA}}{1+b_{1\mu} r_{iA}} \right)^{u_\mu} \left(\frac{a_{2\mu} r_{jA}}{1+b_{2\mu} r_{jA}} \right)^{v_\mu} \left(\frac{a_{3\mu} r_{ij}}{1+b_{3\mu} r_{ij}} \right)^{w_\mu}$
Double exponential[19]	$-\sum\limits_{ij} b e^{-a r_{ij}}$
Gaussian geminal[20]	$\sum\limits_{ij} \sum\limits_{\mu} a_\mu r_{ij}^2$

with which to optimize the wave function. In Table 5.3 we list a number of systems for which b_1 has been optimized for an approximate Hartree-Fock determinant, and the level of correlation energy recovered in the variational energy.

Such Jastrow-type functions have the desirable property that they do not change the nodes created by the determinantal factor, because the correlation functions are positive everywhere. However, their use does have an undesired side effect. Using an SCF determinant and multiplying by Ψ_C causes a global expansion of the electron density.[21] If we assume that the SCF density is relatively accurate, then one needs to rescale the trial function to re-adjust the density. This can be accomplished simply by multiplying by an electron-nucleus Jastrow function. If U for this Jastrow function is given by

$$U = -\sum_{i=1}^{N} \sum_{A=i}^{N_{nuc}} \frac{\lambda_1 r_{iA} + \lambda_2 r_{iA}^2 + \cdots}{1 + \nu_1 r_{iA} + \nu_2 r_{iA}^2 + \cdots}, \tag{5.34}$$

then, as for the electron-electron function, λ_1 is determined by the cusp condition.

For CMO wave functions one can optimize the Jastrow parameters, the molecular orbital coefficients, and the atomic orbital exponents. Clearly, practical limitations

Figure 5.2: Dependence of the linear Padé-Jastrow function on r_{ij}. The coefficient a_1 is set to 0.5 to satisfy the electron-electron cusp condition.

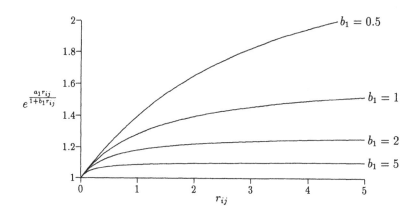

will be reached for very large systems, but such optimization has been done for several systems. Use of a three-parameter correlation function[19] for Li_2 led to 55% of the correlation energy. A 21 parameter function, that included all combinations of r_i, r_j, and r_{ij} to fourth order,[15] including electron-electron-nucleus terms, obtained essentially all the correlation energy for two-electron systems, 99% of the correlation energy for Be, and 86% of the correlation energy for Ne. Use of the Boys-Handy form (shown in Table 5.2) including electron-electron-nucleus terms, gave 68% to 100% of the correlation energy (see Table 5.4). These results are impressive, and show that VMC can be an attractive quantum chemistry technique in its own right.

Exercises

1. Evaluate the local energy of the H atom using trial functions consisting of (a) a

Table 5.3: Correlation energies for selected molecules and optimized values of b_1 in the linear Padé-Jastrow function. The full wave function is the correlation function multiplied by an approximate Hartree-Fock determinant.[22]

System	b_1	%CE
LiH	1.0	21(4)
Li_2	1.0	40(2)
BeH	1.0	32(3)
BH	1.6	27(3)
CH	2.1	19(4)
C_2	2.1	15(3)
NH	2.5	19(4)
N_2	3.2	14(9)
OH	3.7	16(5)
FH	3.6	15(4)

single Slater-type function, e^{-ar}, and (b) a Gaussian-type function, e^{-ar^2}. Plot these two local energies as a function of r for $a = 1.1$. Which of the two would you recommend as a trial function?

2. Starting with the DMC program written for the exercises of Ch. 3, add a linear Jastrow function to the trial function. What is the variance reduction achieved for a fixed amount of computation? How do these variances compare to the variance in a simulation that does not employ a trial function?

Suggestions for Further Reading

1. H. F. Schaefer, III, *The Electronic Structure of Atoms and Molecules. A Survey of Rigorous Quantum Mechanical Results* (Addison-Wesley, Reading, 1972).

2. A. Szabo and N. S. Ostlund, *Modern Quantum Chemistry* (MacMillan Publishing Co., New York, 1982).

Table 5.4: Percent of atomic correlation energy recovered in a VMC calculation with a correlated molecular orbital product wave function. The wave function consists of a Hartree-Fock determinant and a 17-term Boys-Handy function[18] in the form of Eq. 5.32.

System	%CE
He	100(2)
Li	97(7)
Be	68(6)
B	69(5)
C	72(4)
N	77(3)
O	80(3)
F	82(2)
Ne	85(2)

3. A. C. Hurley, *Electron Correlation in Small Molecules* (Academic, New York, 1976).

References

1. See, for example, M. Weissbluth, *Atoms and Molecules*, 570-72 (Academic Press, New York, 1978).

2. H. M. James and A. S. Coolidge, "Criteria of Goodness for Approximate Wave Functions," *Physical Review* **51**, 860-63 (1937).

3. A. A. Frost, R. E. Kellogg and E. A. Curtis, "Local Energy Method in Electronic Energy Calculations," *Reviews of Modern Physics* **32**, 313-17 (1960).

4. K. McDowell, "Assessing the Quality of a Wavefunction Using Quantum Monte Carlo," *International Journal of Quantum Chemistry: Quantum Chemistry Symposium* **15**, 177-81 (1981).

5. T. Kato, "On the Eigenfunctions of Many-Particle Systems in Quantum Mechanics," *Communications on Pure and Applied Mathematics* **10**, 151-77 (1957).

6. C. R. Meyers, C. J. Umrigar, J. P. Sethna and J. D. Morgan, "The Fock Expansion, Kato's Cusp Conditions, and the Exponential Ansatz," *Physical Review A* **44**, 5537-46 (1991).

7. E. A. Hylleraas, "Neue Berechnung der Energie des Heliums in Grand-austande, sowie des tiefsten terms von ortho-Helium," *Zeishrift für Physik* **54**, 347-66 (1929).

8. C. L. Pekeris, "Ground State of Two-Electron Atoms," *Physical Review* **112**, 1649-58 (1958).

9. H. Conroy, "Molecular Schrödinger Equation. I. One-Electron Solutions," *Journal of Chemical Physics* **41**, 1327-31 (1964).

10. H. Conroy, "Molecular Schrödinger Equation. IV. Results for One- and Two-Electron Systems," *Journal of Chemical Physics* **41**, 1341-51 (1964).

11. Y. K. Ho, "Improved Hylleraas Calculations for Ground State Energies of Lithium Iso-Electronic Sequence," *International Journal of Quantum Chemistry* **20**, 1077-82 (1981).

12. K. Frankowski and C. L. Pekeris, "Logarithmic Terms in the Wave Functions of the Ground State of Two Electron Atoms," *Physical Review* **146**, 46-49 (1966).

13. J. D. Baker, J. D. Morgan, D. E. Freund and R. N. Hill, "Radius of Convergence and Analytic Behavior of the $1/Z$ Expansion," *Physical Review A* **43**, 1247-73 (1990).

14. A. C. Hurley, *Introduction to the Electronic Theory of Small Molecules* (Academic, New York, 1976).

15. C. J. Umrigar, K. G. Wilson and J. W. Wilkins, "Optimized Trial Wave Functions for Quantum Monte Carlo Calculations," *Physical Review Letters* **60**, 1719-22 (1988).

16. R. J. Jastrow, "Many-Body Problem with Strong Forces," *Physical Review* **98**, 1479-84 (1955).

17. P. J. Reynolds, D. M. Ceperley, B. J. Alder and W. A. Lester, Jr., "Fixed-Node Quantum Monte Carlo for Molecules," *Journal of Chemical Physics* **77**, 5593-603 (1982).

18. J. W. Moskowitz and K. E. Schmidt, "Correlated Monte Carlo Wave Functions for Some Cations and Anions of the First Row Atoms," *Journal of Chemical Physics* **97**, 3382-85 (1992).

19. S.-Y. Huang, Z. Sun and W. A. Lester, Jr., "Optimized Trial Functions for Quantum Monte Carlo," *Journal of Chemical Physics* **92**, 597-602 (1990).

20. For a recent summary of past Gaussian geminal work see, P. W. Kozlowski and L. Adamowicz, "An Effective Method for Generating Nonadiabatic Many-Body Wave Function Using Explicitly Correlated Gaussian-Type Functions," *Journal of Chemical Physics* **95**, 6681-98 (1991).

21. R. N. Barnett, P. J. Reynolds and W. A. Lester, Jr., "H + H_2 Reaction Barrier: A Fixed-Node Quantum Monte Carlo Study," *Journal of Chemical Physics* **82**, 2700-7 (1985).

22. R. P. Subramaniam, M. A. Lee, K. E. Schmidt and J. W. Moskowitz, "Quantum Simulation of the Electronic Structure of Diatomic Molecules," *Journal of Chemical Physics* **97**, 2600-9 (1988).

Chapter 6

Excited States

Excited states, both vibrational and electronic, play a crucial role in the physics and chemistry of atoms, molecules and condensed matter. In Ch. 5 we saw that the Fermion ground state is itself an excited state — it is the lowest antisymmetric state of a system. Our goal in this chapter is to extend the Monte Carlo methods presented previously for the ground state to higher excited states.

To begin, we write the Monte Carlo solution $\phi(\mathbf{x}, \tau)$ in the now-familiar eigenfunction expansion form,

$$\phi(\mathbf{x}, \tau) = \sum_{i=0}^{\infty} C_i \Phi_i(\mathbf{x}) e^{-(E_i - E_T)\tau}. \tag{6.1}$$

The coefficients C_i of Eq. 6.1 are the overlap integrals of $\phi(\mathbf{x}, 0)$ with the eigenfunctions Φ_i. The function $\phi(\mathbf{x}, \tau)$ will be dominated at large time by the lowest energy solution, Φ_0. If however the coefficient C_0 is zero, i.e. $\phi(\mathbf{x}, 0)$ has no overlap with the ground state, then $\phi(\mathbf{x}, 0)$ would converge to the next-lowest energy state, Φ_1.

With the incorporation of importance sampling, one can make use of both the fixed-node approximation and importance sampling to project out the ground state. Such an approach was implicitly used earlier to project out low-lying Bose states to obtain the Fermion ground state. In general, to eliminate Φ_0 from the trial function Ψ in this manner, Φ_0 must already be known, or else Φ_0 and Ψ must be orthogonal by symmetry. By choosing an antisymmetric Ψ one eliminates *all* symmetric components from $f(\mathbf{x}, \tau)$, but one must impose further constraints on Ψ to obtain excited Fermion states.

In this chapter we present several methods for obtaining excited-state energies. The treatment of other excited-state properties is left to Ch. 7. In Sec. 6.1 we describe ways of exploiting the imaginary-time dependence of the simulation to extract excited-state energies. Section 6.2 deals briefly with the issue of explicit orthogonalization to lower states. Section 6.3 discusses the use of the fixed-node method to obtain excited states. Finally, in Sec. 6.4 we present matrix formulations that can determine the energies of many excited states using an imaginary-time-dependent diagonalization procedure.

6.1   Transforming Energy Decay Curves

From the form of Eq. 6.1 it is evident that the imaginary time-dependence of the Monte Carlo solution contains information about the energy differences $E_i - E_T$. For example, the time-dependence of the energy estimate is given by

$$
\begin{aligned}
E(\tau) \equiv \frac{\int f(\mathbf{x}, \tau) E_L(\mathbf{x}) \, d\mathbf{x}}{\int f(\mathbf{x}, \tau) \, d\mathbf{x}} \;&=\; \frac{\int \phi(\mathbf{x}, \tau) \mathcal{H} \Psi(\mathbf{x}) \, d\mathbf{x}}{\int \phi(\mathbf{x}, \tau) \Psi(\mathbf{x}) \, d\mathbf{x}} \\
&=\; \frac{\sum_{i=0}^{\infty} C_i a_i e^{-(E_i - E_T)\tau} E_i}{\sum_{i=0}^{\infty} C_i a_i e^{-(E_i - E_T)\tau}}.
\end{aligned} \tag{6.2}
$$

Therefore by computing the energy estimate as a function of imaginary time, we can in principle extract these energies. In practice however, obtaining the energies from Eq. 6.2 would be a most difficult task, especially when the statistical noise of the Monte Carlo simulation is non-negligible.

As an alternative, one can devise an estimator specifically to measure the time-dependence. Instead of Eq. 6.2 for the energy, consider the expectation value of the

Green's function, namely,

$$I(\tau) = \langle \Psi | e^{-(\mathcal{H}-E_T)\tau} | \Psi \rangle. \tag{6.3}$$

Equation 6.3 can be sampled from a random walk by evaluating

$$I(\tau) = \langle W(\tau) \rangle_{\psi^2} \tag{6.4}$$

where W is the cumulative branching weight used in pure DMC, i.e.

$$W(\tau) \equiv \prod_{k=0}^{N} e^{-[E_L(\mathbf{x}(k\tau))-E_T]\tau}, \tag{6.5}$$

which is essentially the total population. If we insert a complete set of eigenstates into Eq. 6.3, we see that the time behavior of $I(\tau)$ is

$$I(\tau) = \sum_i a_i^2 e^{-(E_i-E_T)\tau}. \tag{6.6}$$

This is much simpler than Eq. 6.2. The exponential energy-dependence of Eq. 6.6 can be extracted by standard curve fitting methods.

A more sophisticated method involves writing $I(\tau)$ in an integral representation,[1]

$$I(\tau) = \int\limits_{-\infty}^{+\infty} c(E)e^{-\tau E}dE, \tag{6.7}$$

where $c(E)$ is the "energy spectral overlap function,"

$$c(E) = \sum_{i=0}^{\infty} \delta(E - E_i + E_T)\langle \Psi | \Phi_i \rangle^2. \tag{6.8}$$

The substitution of Eq. 6.8 into Eq. 6.7 returns us to Eq. 6.6. Typical $I(\tau)$ and $c(E)$ curves are shown in Fig. 6.1. The E_i may thus be extracted from $I(\tau)$ by using inverse Laplace transform methods to obtain $c(E)$. The computation of $c(E)$ thus provides the excited-state energies and overlaps. However the inversion of Eq. 6.7 is an ill-conditioned problem which requires special statistical methods, such as Bayesian analysis and maximum entropy techniques[1] to extract useful information from the noise.

Figure 6.1: Examples of the correlation function $I(\tau)$ and the spectral overlap function $c(E)$.

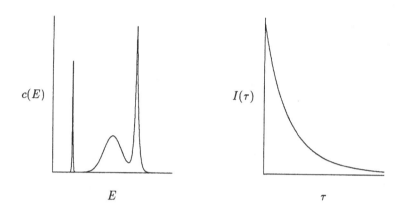

6.2   Explicit Orthogonalization

An alternative approach for obtaining excited state information is available if one already has a fairly accurate representation of the exact ground state, Φ_0, of a system. If Φ_0 is known analytically, we can construct a state orthogonal to it fairly easily. If instead Φ_0 is obtained from a quantum Monte Carlo calculation, where it is represented as the density of walkers, how do we explicitly create a Monte Carlo state to represent Φ_1?

One method[2] for doing so is to assign each walker a weight of $\omega_k = \pm 1$, where the signs are chosen to enforce the constraint,

$$D_0 \equiv \sum_{k=1}^{M} \Phi_0(\mathbf{x}_k)\omega_k = 0, \tag{6.9}$$

upon an ensemble of M walkers. The initial weights are generally chosen to be equal to the sign of the excited state wave function at $\mathbf{x}_k$, though this is only known approximately, e.g. via a trial function. Proper enforcement of this constraint builds in orthogonality to Φ_0 by forcing the overlap of the excited state distribution (with signs) with the known ground state to vanish. This orthogonality is accomplished by initially establishing an ensemble satisfying $|D_0| \leq \epsilon$ (where ϵ is a chosen cutoff value) based upon some *a priori* knowledge of the excited state, such as the number of nodal volumes or its symmetry. If D_0 exceeds this value as the walk proceeds, positively or negatively weighted walkers are created or destroyed systematically to restore the constraint. In this way the state Φ_1 can be extracted (in a Monte Carlo sense) from the distribution of walkers.

Once we have a representation of Φ_1, this process can be continued to higher excited states by adding the constraint

$$D_1 \equiv \sum_{k=1}^{M} \Phi_1(\mathbf{x}_k)\omega_k = 0. \tag{6.10}$$

In principle this process may be continued for as many states as desired, but, because each state must be determined successively by Monte Carlo, the statistical errors in the function Φ_n will increase rapidly as n increases. Nevertheless, this method and other more sophisticated variants of this method have been applied successfully to model systems.

6.3   Fixed-Node Method

We have encountered the fixed-node approximation in the context of obtaining the antisymmetric, Fermion ground state. Fixing the nodes, is a general, stable method for constraining a QMC simulation. If the exact nodes are known then the exact excited state will result. Even if the exact nodes are not known, the spatial[3] or spin symmetries[6] can be built into the trial function so that antisymmetric states can

Figure 6.2: Effect of nodal misplacement for the particle in a box. The trial node is displaced by an amount δ from the exact node located at the origin.

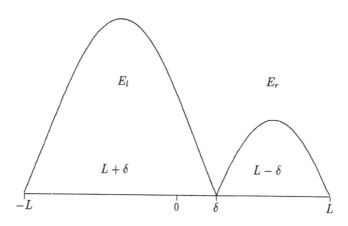

be separated from symmetric states, electron-spin multiplets can be separated, and irreducible representations of symmetry point groups can be distinguished.

In Ch. 3 we showed that the fixed-node approximation is variational with respect to the exact ground state energy. This result can be generalized to show that any state which is the *lowest* energy state of a given symmetry will be a variational bound to the exact energy for that state. However, for states that have the *same* symmetry as lower states, the resulting energy will be given by a linear combination of energies that include the lower state. If the trial function is a good approximation to the state of interest, the other coefficients will be small. Specifically, if Ψ_i is the trial function for state i,

$$\tilde{E}_i = \left\langle \frac{\mathcal{H}\Psi_i}{\Psi_i} \right\rangle_{f_\infty} = \sum_{j=0}^{\infty} b_{ij} E_j. \qquad (6.11)$$

The coefficients b_{ij} are related to the overlap integral $\langle \Psi_i | \Phi_j \rangle$. Ideally $b_{ij} = \delta_{ij}$;

however, mixing of energies can occur as in Eq. 6.11 because the incorrect placement of nodes may lower the energy in some nodal volumes. Consider the case of the first excited state of the particle in a box with walls at $\pm L$ as shown in Fig. 6.2. The exact node is at the origin. In the fixed-node method, the two nodal regions are simulated separately since no walkers may cross a node. The exact energy for a box of width a is given by,

$$E_n(a) = \frac{\hbar^2 \pi^2 (n+1)^2}{2ma^2} \qquad\qquad n = 0, 1, \cdots. \qquad (6.12)$$

In this case the nodeless ground-state $(n = 0)$ energy is obtained by setting $a = 2L$. The energy of the first excited state is given by $n = 1$ and $a = 2L$, which is the same as the ground-state energy of two equal nodal regions, i.e. $n = 0$ and $a = L$. However, if the node is displaced to the right by δ as indicated in Fig. 6.2, then the ground state energy on the right becomes

$$E_r = \frac{\hbar^2 \pi^2}{2m(L-\delta)^2}, \qquad (6.13)$$

and on the left it becomes

$$E_l = \frac{\hbar^2 \pi^2}{2m(L+\delta)^2}. \qquad (6.14)$$

Clearly $E_l < E_r$. Because the branching is determined by a rate constant which depends on the local energy, the population of walkers in the right-hand region will decrease exponentially with respect to that in the left-hand region, and the long-time solution will yield only E_l. Thus, the QMC energy will not be variationally bounded by any excited state energy E_i with $i > 0$, although it is always bounded by E_0. Nevertheless, the mixing is often small, and a good approximation to E_i can be obtained.

Several representative excited state calculations using this method are given in Table 6.1. It appears that in general *ab initio* trial wave functions have the correct nodal structure to obtain good estimates for excited states, even when no variational

Table 6.1: Excited-state transition energies (in eV) calculated by fixed-node DMC.

System	QMC	Experiment	Ref.
$H_2(X\ ^1\Sigma_g^+ \rightarrow B\ ^1\Sigma_u^+)$	20.621(2)	20.619	3
$H_2(X\ ^1\Sigma_g^+ \rightarrow E\ ^1\Sigma_g^+)$	22.927(2)	22.924	3
$He(1s^2 \rightarrow 1s2s)$	20.621(2)	20.619	4
$He(1s^2 \rightarrow 1s3s)$	22.927(2)	22.924	4
$Li(^2S \rightarrow\ ^2P)$	1.844(9)	1.848	5
$CH_2(^3B_1 \rightarrow\ ^1A_1)$	8.9(2.2)	9.1	6

bound exists. For example, in H_2 the B state is the lowest state of Σ_u symmetry, and the result is exact within the error bars as expected. For the E state, which has the same symmetry as the ground state, the result is somewhat worse, although it is still within the error bar of the exact result.

One aspect of computing excited states is that the Hartree-Fock determinant rarely is an adequate trial function. In particular, open shell singlet states such as the B and E states, cannot be properly specified with only a single determinant. To explain the need for more than a single determinant, let us examine the trial functions for H_2 more closely. The occupied molecular orbitals are $1\sigma_g$, $1\sigma_u$, and $2\sigma_g$ (the latter only for the E state). If we wanted a triplet state it could be properly written, e.g., as $\det|1\sigma_g(1)\alpha(1)\ 1\sigma_u(2)\alpha(2)|$, which poses no problem in QMC. However, an open-shell singlet state must be written as,

$$1\sigma_g(1)\alpha(1)1\sigma_u(2)\beta(2) - 1\sigma_g(1)\beta(1)1\sigma_u(2)\alpha(2). \tag{6.15}$$

This wave function presents a problem in QMC because we have no way to represent the spin functions. Rather, in QMC one separates the α and β determinants and writes their product. The resulting singlet state, $1\sigma_g(1)\alpha(1)1\sigma_u(2)\beta(2)$, is not antisymmetric. Hence, the correct trial function, Eq. 6.15 can only be obtained as the difference of two determinants.

6.4   Concurrent Evaluation of Many States

In CI methods, the wave function is far more complex than those we have been using here (cf. Sec. 5.3.5). However, CI provides the excited-state energies and wave functions by a straightforward diagonalization of the Hamiltonian matrix $\langle \Psi_i | \mathcal{H} | \Psi_j \rangle$. Can such a *matrix* procedure, which provides information about many excited states at once, be implemented in QMC? In this section we discuss one method to do so.

To begin, let us generalize Eq. 6.1 to a manifold of N QMC states,[7] i.e.,

$$\phi_i(\mathbf{x}, \tau) = e^{-\tau \mathcal{H}} | \Psi_i(\mathbf{x}) \rangle = \sum_{k=0}^{\infty} C_{ik} \Phi_k e^{-E_k \tau}. \tag{6.16}$$

(Note we have dropped the use of the energy shift E_T for later convenience.) B These states form a QMC *basis set* of evolving states produced from N trial functions Ψ_i. Now consider the overlap and Hamiltonian matrices in this imaginary time dependent basis, i.e.

$$S_{ij}(\tau) \equiv \langle \phi_i(\mathbf{x}, \tau) | \phi_j(\mathbf{x}, \tau) \rangle \tag{6.17}$$

$$H_{ij}(\tau) \equiv \langle \phi_i(\mathbf{x}, \tau) | \mathcal{H} | \phi_j(\mathbf{x}, \tau) \rangle. \tag{6.18}$$

With these matrices one can compose a generalized eigenvalue problem $\mathbf{H} - \lambda \mathbf{S} = 0$. The eigenvalues λ, i.e. the solutions of the matrix equation

$$\det | H_{ij}(\tau) - \lambda S_{ij}(\tau) | = 0, \tag{6.19}$$

will converge to the exact eigenvalues of $\mathcal{H}$ as $\tau \to \infty$. To show this, we expand the determinant in Eq. 6.19 in terms of the eigenvalues of $\mathcal{H}$ (using Eq. 6.16), to obtain

$$\det|H_{ij} - \lambda S_{ij}| = \sum_{k_0 < k_1 < \cdots < k_{N-1}}^{\infty} [E_{k_0} - \lambda]e^{-2\tau E_{k_0}} \cdots [E_{k_{N-1}} - \lambda]e^{-2\tau E_{k_{N-1}}}$$
$$\times \det|C_{0k_0} \cdots C_{N-1k_{N-1}}|^2. \tag{6.20}$$

Each subscript $k_0, \cdots, k_{N-1}$ denotes an index that is to be summed over all the exact states of the system. The full sum is over all possible combinations of the infinite number of exact states taken N (the number of QMC basis functions) at a time, beginning with $k_0 = 0$, $k_1 = 1$, $\cdots$, $k_{N-1} = N - 1$. Eq. 6.20 is the result of tedious manipulations of Eq. 6.17, Eq. 6.18, and Eq. 6.19. To see how it comes about, note that a single element of the matrix can be rearranged to give,

$$H_{ij} - \lambda S_{ij} = \sum_{k=0}^{\infty} C_{ik} C_{jk} e^{-2\tau E_k}(E_k - \lambda). \tag{6.21}$$

A determinant is the signed sum over permutations of the indices of the product of the diagonal elements of a matrix. For Eq. 6.19 the diagonal product is,

$$\prod_{i=0}^{N-1} (H_{ii} - \lambda S_{ii}) = (H_{00} - \lambda S_{00}) \cdots (H_{N-1\,N-1} - \lambda S_{N-1\,N-1})$$
$$= \left(\sum_{k_0=0}^{\infty} C_{0k_0}^2 e^{-2\tau E_{k_0}} (E_{k_0} - \lambda) \right)$$
$$\cdots \left(\sum_{k_{N-1}=0}^{\infty} C_{N-1\,k_{N-1}}^2 e^{-2\tau E_{k_{N-1}}} (E_{k_{N-1}} - \lambda) \right)$$
$$= \prod_{i=0}^{N-1} \sum_{k_i=0}^{\infty} C_{ik_i}^2 (E_{k_i} - \lambda)e^{-2\tau E_{k_i}}. \tag{6.22}$$

By reordering the sums over k's and resumming the coefficients into a determinant, one obtains the desired result.

As is the case with single-state QMC, at large time the only contributor to the sum in Eq. 6.20 will be the first term. All terms containing higher excited states than

$N - 1$ will decay asymptotically with respect to the first term, which yields the exact energy of the N lowest states. That is,

$$\lim_{\tau \to \infty} \det |H_{ij}(\tau) - \lambda S_{ij}(\tau)| = e^{-2\tau(E_0 + E_1 + \cdots + E_{N-1})} [E_0 - \lambda] \cdots [E_{N-1} - \lambda]$$
$$\times \det |C_{00} \cdots C_{N-1 N-1}|^2. \tag{6.23}$$

Making use of this formalism requires a QMC simulation that produces ϕ_0 to ϕ_{N-1} simultaneously. To do this, one can start by rewriting the equations for H_{ij} and S_{ij} in terms of the trial functions $\Psi_i, i = 0, \cdots, N-1$, as

$$S_{ij}(\tau) = \langle \Psi_i | e^{-2\tau \mathcal{H}} | \Psi_j \rangle \tag{6.24}$$

$$H_{ij}(\tau) = \langle \Psi_i | \mathcal{H} e^{-2\tau \mathcal{H}} | \Psi_j \rangle, \tag{6.25}$$

where we have used Eq. 6.16 and factored the time propagator back out of the QMC states. This propagator may be evaluated using the cumulative branching weight previously given in Eq. 6.5.

Since all the $\phi_i(\mathbf{x}, \tau)$ need to be propagated together, a simple guiding function Ψ_G should be used. The guiding function creates the Monte Carlo distribution, and its local energy is used in the weights. Because the distribution created is to be used to evaluate the matrix elements of Eqs. 6.24 and 6.25, it is of primary importance to choose a Ψ_G that has maximal overlap with each Ψ_i. This can be done either with a simple function chosen to overlap all states, such as a single diffuse Gaussian, or with a superposition of states such as,

$$\Psi_G = \sum_i [b_i \Psi_i^m]^{1/m}, \tag{6.26}$$

where b_i and m are chosen to adjust the relative contributions (and hence convergence times) of the various states.

Although H_{ij} and S_{ij} are needed only for $\tau \to \infty$, they are best calculated at a series of times, τ_k, so that convergence can be observed. A VMC walk can be used to create the Ψ_G^2 distribution. At time τ_0 the quantities $F_i \equiv \Psi_i/\Psi_G$ and $E_i \equiv \mathcal{H}\Psi_i/\Psi_G$ are computed for the ensemble and stored. The walk then proceeds to time τ_1, where $W(\tau_1)$ together with the new F_i and E_i are calculated. This process is repeated until the maximum convergence time to be considered, τ_m, is reached. The matrices can then be constructed as

$$
\begin{aligned}
H_{ij}(\tau_k) \;=\; & \frac{1}{4}\left[F_i(\tau_k)W(\tau_k)E_j(0) + F_j(\tau_k)W(\tau_k)E_i(0)\right. \\
& \left. + F_i(0)W(\tau_k)E_j(\tau_k) + F_j(0)W(\tau_k)E_i(\tau_k)\right]
\end{aligned}
\tag{6.27}
$$

and

$$
S_{ij}(\tau_k) = \frac{1}{2}\left[F_i(\tau_k)W(\tau_k)F_j(0) + F_j(\tau_k)W(\tau_k)F_i(0)\right]
\tag{6.28}
$$

where the symmetry in i, j and τ of H_{ij} and S_{ij} has been exploited.

To clarify this method we provide a simple schematic algorithm here.

ALGORITHM 6.1 Matrix QMC determination of excited states.

```
C
C Conduct a VMC walk with guiding function PSIG. Allow ensemble to equilibrate.
C
      Do Istep=1,Nstep
         Do k=1,Nwalkers
C
C Compute initial values of the trial functions PSIi and their local
C energies ELi. Then weight to obtain F and E as indicated in the text.
C Note we compute quantities only every Mt steps and must keep the quantities
C separate. TRIALF computes quantities with respect to the trial functions
C and GUIDEF computes quantities with respect to the guiding function.
C
                  If (MOD(Nstep,Mt) = 0) Then
                     L = Istep/Mt
                     CALL TRIALF(PSIi,ELi)
                     CALL GUIDEF(PSIG,ELG)
                     Do i=1,Nstate
                        F(i,L,k) = PSIi(i)/PSIG(k)
                        E(i,L,k) = ELi(i)/F(i,L,k)
                     EndDo
                  Endif
C
C Move walker as usual then compute the log of the branching weight.
C
                  ...Move walker...
                  WL = EXP(-(ELG - ET)*Tau)
```

```
                Do L=1,Nj
                    W(L,k) = W(L,K) + WL
                EndDo
C
C Once we have reached the desired correlation time Tcorr, sample Hij and Sij.
C LT is the current time-correlation interval, LO is the "zero" of time.
C
                Time = Tau*Istep - Tcorr
                If (Time > 0 and MOD(Time,Mt) = 0) Then
                    LT = Istep/Mt
                    LO = Istep/Mt - Tcorr/Tau/Mt
                    Do i=1,Nstate
                        Do j=1,Nstate
                            H(i,j,k) = H(i,j,k) + 1/4*W(LO,k)*[
                                 E(i,LT,k)*F(j,LO,k) + E(j,LT,k)*F(i,LO,k)
                               + E(i,LO,k)*F(j,LT,k) + E(j,LO,k)*F(i,LT,k) ]
                            S(i,j,k) = S(i,j,k) + 1/2*W(LO,k)*[
                                 F(i,LT,k)*F(j,LO,k) + F(j,LO,k)*F(i,LT,k) ]
                        EndDo
                    EndDo
                EndIf
            End Do
        End Do
C
C At end of walk, find the average of Hij and Sij and diagonalize
C Hij - E Sij. Also diagonalize at every block to estimate the variance
C of the energies.
C
        End
```

As discussed above the energies are extracted as the eigenvalues of $\mathbf{H} - \lambda\mathbf{S} = 0$. However, because the matrix elements are statistical, care must be taken to assure convergence to the proper states. It is not difficult to show that because diagonalization is a non-linear process, noise in the matrix elements will result in biasing the resulting energies. In particular, noise causes the estimate of the lowest energies to be pushed down and the highest energies to be pushed up. Hence the best energies will result from values of H_{ij} and S_{ij} with the smallest variance.

Note that the weights will either vanish or diverge at large time (as in the pure DMC algorithm). Moreover, just as in the released node method, the variance rises exponentially because *all* the QMC basis functions are converging to the *ground state*. Hence, only a *transient* estimate of the energy is obtained, as illustrated in Fig. 6.3. In particular, the higher excited states will be more poorly converged than the lower states. For this reason this method was first applied to Bose systems (such as vibra-

Figure 6.3: Typical behavior of the excited-state energies for matrix QMC.

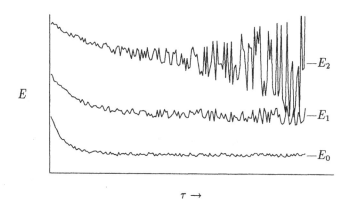

tional states[8]) where the energy gap between ground and excited states is relatively small. One method to stabilize the transient estimate as long as possible, and to reduce any bias as much as possible, is to re-diagonalize $|H_{ij}(\tau_k) - \lambda S_{ij}(\tau_k)|$ at each time interval. The resulting eigenfunctions can be used as new trial functions. In this way the maximum overlap with the exact states can be maintained. Results for the vibrational energies of formaldehyde are given in Table 6.2.

Table 6.2: Excitation energies $E_k - E_0$ for the vibrational states of H_2CO by matrix QMC.[8] Energies are in cm^{-1}.

Symmetry	State	QMC	Vibrational SCF	Experiment
A_1	1	5970(2)[a]		
	2	1520(10)	1464	1500
	3	1756(6)	1792	1746
	4	2253(8)		2327
	5	2457(26)		
	6	2790(23)	2690	2783
	7	3110(60)		3000
	8	3316(60)		3239
B_1	1	1146(3)	1164	1167
	2	2702(9)		2656
	3	2888(25)		2905
	4	3310(30)		
	5	3610(50)		
	6	4000(30)		
	7	4312(40)		
	8	4534(40)		
B_2	1	1228(8)	1239	1249
	2	2796(30)		2719
	3	2830(30)	2836	2843
	4	3045(30)		3000
	5	3490(30)		
	6	4180(50)		
A_2	1	2359(10)		
	2	3910(40)		
	3	3950(40)		
	4	4205(30)		

(a) Ground state energy.

Exercises

1. Show that the fixed-node energy for the first state of a symmetry different than the ground state is an upper bound to the exact excited-state energy.

2. Using the fact that H_{ij} and S_{ij} are symmetric matrices, and that forward and reverse time are equivalent, derive Eqs. 6.27 and 6.28.

3. For a 2×2 determinant show that adding noise to the matrix elements lowers the smallest eigenvalue and raises the largest.

Suggestions for Further Reading

1. J. D. Graybeal, *Molecular Spectroscopy*, (McGraw-Hill, New York, 1988).

2. A. C. Hurley, *Electron Correlation in Small Molecules*, (Academic, New York, 1976).

3. D. M. Hirst, *A Computational Approach to Chemistry*, (Blackwell Scientific, London, 1990).

4. H. F. Schaefer, III, *The Electronic Structure of Atoms and Molecules. A Survey of Rigorous Quantum Mechanical Results*, (Addison-Wesley, Reading, MA, 1972).

5. J. S. Bowers, R. K. Prud'homme and R. S. Farinato, "An Assessment of the Padé-Laplace Method for Transient Electric Birefringence Decay Analysis," *Computers & Chemistry* **16**, 249-59 (1992).

References

1. M. Caffarel and D. M. Ceperley, "A Bayesian Analysis of Green's Function Monte Carlo Correlation Functions," *Journal of Chemical Physics* **97**, 8415-23 (1992).

2. D. F. Coker and R. O. Watts, "Quantum Simulation of Systems with Nodal Surfaces," *Molecular Physics* **58**, 1113-23 (1986).

3. R. M. Grimes, B. L. Hammond, P. J. Reynolds and W. A. Lester, Jr., "Quantum Monte Carlo Approach to Electronically Excited Molecules," *Journal of Chemical Physics* **85**, 4749-50 (1986).

4. P. J. Reynolds, R. N. Barnett, B. L. Hammond and W. A. Lester, Jr., "Molecular Physics and Chemistry Applications of Quantum Monte Carlo," *Journal of Statistical Physics* **43**, 1017-26 (1986).

5. R. N. Barnett, P. J. Reynolds and W. A. Lester, Jr., "Monte Carlo Determination of the Oscillator Strength and Excited State Lifetime for the Li $2^2S \rightarrow 2^2P$ Transition," *International Journal of Quantum Chemistry* **42**, 837-47 (1992).

6. P. J. Reynolds, M. Dupuis and W. A. Lester, Jr., "Quantum Monte Carlo Calculation of the Singlet-Triplet Splitting in Methylene," *Journal of Chemical Physics* **82**, 1983-90 (1985).

7. D. M. Ceperley and B. Bernu, "The Calculation of Excited-State Properties with Quantum Monte Carlo," *Journal of Chemical Physics* **89**, 6316-28 (1988).

8. B. Bernu, D. M. Ceperley and W. A. Lester, Jr., "The Calculation of Excited States with Quantum Monte Carlo. II. Vibrational Excited States," *Journal of Chemical Physics* **93**, 552-61 (1990).

Chapter 7

Electronic Properties

In the preceding chapters we described how to use Monte Carlo to obtain the electronic energy of an atomic or molecular system. In particular, we learned that one can compute the energy as the average value of the local energy E_L with respect to the *mixed distribution*, $\Phi_0\Psi$. As we saw in Ch. 3, this produces the ground state energy,

$$E_{\text{QMC}} = \frac{\int \Phi_0 \mathcal{H} \Psi \, \mathrm{d}\mathbf{x}}{\int \Psi \Phi_0 \, \mathrm{d}\mathbf{x}} = E_0. \tag{7.1}$$

The crucial step in deriving Eq. 7.1 requires the equality of $\langle \Phi_0 | \mathcal{H} | \Psi \rangle$ with $\langle \Psi | \mathcal{H} | \Phi_0 \rangle$, together with the eigenvalue equation, $\mathcal{H}\Phi_0 = E_0\Phi_0$. However, for an operator that does not obey an eigenvalue equation with respect to Φ_0, i.e. for one that does not commute with the Hamiltonian, the mixed estimator does not yield the correct expectation value. For such an operator, $\mathcal{A}$, the expectation value A is

$$A \equiv \frac{\int \Phi_0 A \Phi_0 \, \mathrm{d}\mathbf{x}}{\int \Phi_0^2 \, \mathrm{d}\mathbf{x}} \neq \frac{\int \Phi_0 A \Psi \, \mathrm{d}\mathbf{x}}{\int \Phi_0 \Psi \, \mathrm{d}\mathbf{x}}. \tag{7.2}$$

Therefore, to use QMC to estimate A we need methods to sample from the "pure" Φ_0^2 distribution.

The form of $\mathcal{A}$ is important. Expectation values of coordinate operators, which lead to quantities such as the dipole, quadrupole, and higher moments, are straightforward to compute by QMC. However other operators, such as electric and magnetic

199

fields, which perturb the wave function, necessitate the computation of the response of the wave function to the perturbation. But in QMC one does not have an explicit expression for the wave function. In those cases, one must resort either to finite difference or analytic derivative methods appropriate to QMC, to determine the response.

In this chapter we will discuss the class of multiplicative coordinate operators. For these operators, the various methods to sample from the pure Φ_0^2 distribution are all that is required. These will be presented, starting with approximate methods and proceeding to exact ones. The class of operators that requires the response of the QMC wave function will be discussed separately in Ch. 8.

7.1 VMC Properties

VMC is the simplest Monte Carlo method for computing properties. The accuracy of computed properties is determined by the choice of Ψ. Specifically, the average value of any function of the coordinates $A \equiv A(\mathbf{x})$, may be evaluated as

$$A_v \equiv \frac{\int \Psi(\mathbf{x}) A(\mathbf{x}) \Psi(\mathbf{x}) \, d\mathbf{x}}{\int \Psi^2(\mathbf{x}) \, d\mathbf{x}} = \lim_{M \to \infty} \frac{1}{M} \sum_{k=1}^{M} A(\mathbf{X}_k), \qquad (7.3)$$

where A_v is known as the variational estimator of A. When using a separate guiding function, Ψ_G, this expression becomes

$$A_v = \lim_{M \to \infty} \left[\frac{\sum_{k=1}^{M} A(\mathbf{Y}_k) w^2(\mathbf{Y}_k)}{\sum_{k=1}^{M} w^2(\mathbf{Y}_k)} \right]. \qquad (7.4)$$

In these equations $\{\mathbf{X}\}$ represents a set of configurations drawn from the distribution Ψ^2, $\{\mathbf{Y}\}$ are drawn from Ψ_G^2, and the function $w = \Psi/\Psi_G$. The latter form, Eq. 7.4, has the advantage that a relatively simple Ψ_G can be chosen to generate the random walk. As discussed in Ch. 2, the sample points $\{\mathbf{Y}\}$ at which A is evaluated can then be chosen as a subset of the full random walk, e.g. every n-th step. This reduces the serial correlation between sample points. Because the trial function Ψ would need to be evaluated only every n steps, a relatively complex and more accurate form could be used with only modest additional computational effort. A balance must be established empirically for each (Ψ_G, Ψ) pair, to find the optimum n that weighs the gain in sampling efficiency due to decorrelation of the sample against the loss of statistical precision due to sampling n-fold fewer points.

Any multiplicative operator can be evaluated by Eq. 7.3 or Eq. 7.4 in VMC, because the wave function is known explicitly. Moreover, because derivatives of a known function may readily be computed, there is also no difficulty in sampling non-multiplicative operators. Several properties are listed in Table 7.1. Perhaps the most novel of those listed is an estimate the relativistic energy corrections obtained by measuring averages of first-order perturbation expressions.

Despite its potential, relatively little work has been done with VMC properties. With recent improvements in the ability to generate both correlation functions and fully optimized VMC trial functions, we anticipate that VMC will play a larger role in the future.

7.2   Approximate Φ_0^2 Estimators

Before turning to exact methods of sampling from Φ_0^2, let us examine approximate methods. These are useful because they are both simpler and less costly. Their value will depend on the quality of the results obtainable. To begin, let us define the mixed estimator of $A(\mathbf{x})$ as,

$$A_m \equiv \frac{\int \Phi_0 A(\mathbf{x}) \Psi \, d\mathbf{x}}{\int \Phi_0 \Psi \, d\mathbf{x}}, \tag{7.5}$$

to distinguish it from the pure estimator, which we write in the form,

$$A_p \equiv \frac{\int \Phi_0 A(\mathbf{x}) \Phi_0 \, d\mathbf{x}}{\int \Phi_0^2 \, d\mathbf{x}}. \tag{7.6}$$

Clearly both the mixed and the earlier defined variational estimators are approximations to the pure expectation value. However, A_m can be combined with A_v to provide a better approximation to the pure estimator. To understand the relationship among A_v, A_m and A_p, we can define the difference $\Delta \equiv \Phi_0 - \Psi$, and rewrite A_m in a Taylor series in Δ. This leads to

$$A_m = A_p + \int \Phi_0 (A_p - A(\mathbf{x})) \, \Delta \, d\mathbf{x} + O(\Delta^2), \qquad (7.7)$$

and

$$A_v = A_p + 2 \int \Phi_0 (A_p - A(\mathbf{x})) \, \Delta \, d\mathbf{x} + O(\Delta^2). \qquad (7.8)$$

Both A_v and A_m have errors which are first-order in Δ. If we eliminate the common first-order term between Eq. 7.7 and Eq. 7.8, we obtain a second-order approximation[3] to A_p, namely

$$A_p \approx \tilde{A}_p \equiv 2A_m - A_v = A_p + O(\Delta^2). \qquad (7.9)$$

Equation 7.9 has been used extensively to calculate QMC properties. In Table 7.2 we present representative results using Eq. 7.9, and compare them with A_v, A_m, A_p (see below) and experiment. One can see that Eq. 7.9 does in fact provide considerably improved results over both A_v and A_m, but, as expected, is not as accurate as A_p.

Equation 7.9 seems to indicate that two separate calculations, one for A_v and one for A_m, must be done. The resulting error for the approximate value would therefore be larger than the error in either value alone. However, because VMC is usually much faster than QMC, in practice the error and computation time of $\tilde{A}_p$ is not much larger than that for A_m. It also is possible to calculate both A_v and A_m in a single walk using pure DMC.

Table 7.1: Selected molecular properties (in a.u.) calculated by VMC.

System	Property	QMC	Exact
H_2[1]	$\langle r_{12} \rangle$	2.168(2)	2.1689
	$\langle r_{12}^2 \rangle$	5.629(9)	5.6324
	$\langle r_{12}^{-1} \rangle$	0.5882(9)	0.58737
	$\langle r_1 \rangle$	1.548(2)	1.5488
	$\langle r_1^2 \rangle$	3.031(6)	3.0364
	quadrupole moment Q	0.473(5)	0.4568
	hexadecapole moment	0.27(4)	0.2826
	polarizability $\alpha_\parallel$	5.57(1)	6.380
	polarizability $\alpha_\perp$	4.11(1)	4.578
LiH[2]	Relativistic energy correction[a]	$-7.3(2) \times 10^{-4}$	-7.97×10^{-4}

(a) The "exact" column quotes the relativistic Hartree-Fock result.

7.3 Rigorous Sampling of Φ_0^2

Several methods exist to sample from the distribution Φ_0^2. We begin our discussion with the future-walking method. In this approach one samples f_∞ as in ordinary QMC, but one also obtains the weight Φ_0/Ψ. A second approach, the time correlation method, samples from the variational distribution Ψ^2, and obtains separately the weight Φ_0^2/Ψ^2. This latter approach is similar to pure DMC, and is presented below, following a discussion on future walking. We conclude this section with a look at the connections between these two algorithms.

7.3.1 Future walking

The future-walking method may be coupled with any form of QMC which leads to a mixed distribution, such as DMC. In essence, one computes A_p by multiplicatively

Table 7.2: Comparison of A_v, A_m, $\tilde{A}_p$ and A_p estimators for H_2 and N_2.[4] Units are a.u.

Method	$\frac{1}{2}(\langle x^2 \rangle + \langle y^2 \rangle)$	$\langle z^2 \rangle$	$\langle r^2 \rangle$	Q	N$_2$ Q
SCF	0.7768	1.020	2.574	0.66	-1.29
A_v	0.772(1)	1.078(2)	2.621(3)	0.49(1)	$-2.19(4)$
A_m	0.767(2)	1.047(4)	2.580(9)	0.56(2)	$-1.8(1)$
$\tilde{A}_p$	0.763(5)	1.016(9)	2.539(18)	0.63(5)	$-1.4(2)$
A_p	0.764(3)	1.025(5)	2.552(10)	0.61(3)	
Exact	0.7617	1.023	2.546	0.61	$-1.4(1)$

including a weight representing the ratio Φ_0/Ψ in Eq. 7.5. This weight, which we shall refer to as the future-walking weight, is obtained[5] from the asymptotic number of descendents of a single walker.

If the initial QMC distribution, $f(\mathbf{x}, 0) \equiv \phi(\mathbf{x}, 0)\Psi(\mathbf{x})$, is merely a single walker at position $\mathbf{X}$, then that distribution may be represented by a delta function in electronic coordinate space, i.e. $f(\mathbf{x}, 0) = \delta(\mathbf{x} - \mathbf{X})$. By expanding $\phi(\mathbf{x}, 0)$ in eigenfunctions of $\mathcal{H}$, we may represent the delta function as

$$\delta(\mathbf{x} - \mathbf{X}) = \Psi(\mathbf{x}) \sum_{j=0}^{\infty} c_j(\mathbf{X}) \Phi_j(\mathbf{x}). \tag{7.10}$$

The coefficients c_j may be obtained by multiplying both sides of Eq. 7.10 by Φ_i/Ψ, and integrating over $\mathbf{x}$. This manipulation yields the coefficient,

$$c_i(\mathbf{X}) = \int \delta(\mathbf{x} - \mathbf{X}) \frac{\Phi_i(\mathbf{x})}{\Psi(\mathbf{x})} d\mathbf{x} = \frac{\Phi_i(\mathbf{X})}{\Psi(\mathbf{X})}. \tag{7.11}$$

Thus $c_0(\mathbf{X})$ is the desired weight we need for the ground state. But what quantity can we sample which will give us c_0?

As we already have seen, during the course of the QMC walk, repeated application of the Green's function causes the ground state to grow exponentially with respect to the excited states, leading to the asymptotic imaginary-time behavior

$$f(\mathbf{x}, \tau \to \infty) = c_0 e^{-(E_0 - E_T)\tau} \Psi(\mathbf{x}) \Phi_0(\mathbf{x}). \tag{7.12}$$

The *total* number of walkers (in branching algorithms) or total weight of walkers (in general) at time τ is simply the integral of the density of walkers, i.e.

$$P(\tau) = \int f(\mathbf{x}, \tau) \, d\mathbf{x}. \tag{7.13}$$

If we define $P_\infty(\mathbf{X})$ to be the asymptotic population of walkers descended from $\mathbf{X}$, we find

$$P_\infty(\mathbf{X}) = \int c_0(\mathbf{X}) e^{-(E_0 - E_T)\tau} \Psi(\mathbf{x}) \Phi_0(\mathbf{x}) \, d\mathbf{x} = \frac{\Phi_0(\mathbf{X})}{\Psi(\mathbf{X})} e^{-(E_0 - E_T)\tau} \langle \Psi | \Phi_0 \rangle. \tag{7.14}$$

$P_\infty(\mathbf{X})$ contains the ratio we need, and can be used directly as a weight to convert Eq. 7.5 to to yield the pure average of Eq. 7.6, i.e.

$$A_p = \frac{\int \Phi_0(\mathbf{x}) A(\mathbf{x}) P_\infty(\mathbf{x}) \Psi(\mathbf{x}) \, d\mathbf{x}}{\int \Phi_0(\mathbf{x}) P_\infty(\mathbf{x}) \Psi(\mathbf{x}) \, d\mathbf{x}} = \lim_{M \to \infty} \frac{\sum_{k=1}^{M} A(\mathbf{X}_k) P_\infty(\mathbf{X}_k)}{\sum_{k=1}^{M} P_\infty(\mathbf{X}_k)}, \tag{7.15}$$

where $\{\mathbf{X}\}$ is drawn from $f_\infty = \Phi_0 \Psi$. The space-independent terms in Eq. 7.14 cancel between the numerator and denominator of Eq. 7.15.

In practice, each walker at each step of the walk will become the parent in Eq. 7.14. Assuming we start with an initial ensemble of N walkers, each of these walkers will become the root of a "family tree" of daughter walkers. However, one cannot compute the average in Eq. 7.15 until P_∞ has been determined — which requires continuing the walk from $\mathbf{X}_k$ into the "future" for at least a time s, where s is the convergence time to reach the asymptotic population. For a time step $\delta\tau$, this requires $m_s = s/\delta\tau$ steps. As a consequence, the values of $A(\mathbf{X}(t))$ together with the family trees or parentage tables must be kept for m_s time steps. The pure averages, A_p, listed in Table 7.2 were obtained using this method.

To illustrate how such an algorithm might work, let us outline one implementation of this method, which is described in detail in Ref. 5. In this algorithm, the parentage is determined by placing each parent walker at the origin of a circle. Each descendent is placed at a radial coordinate which is the walk time, t. In addition, each walker is assigned an angular coordinate θ and range, δ, in which all of its descendents will be placed. Hence, knowing t, θ and δ for a given walker specifies the location of all that walkers' descendents. Specifically, the asymptotic population is the number of walkers found at radius $t + s$ between θ and $\theta + \delta$ in the circle. Algorithm 7.1 outlines this.

ALGORITHM 7.1 Future walking algorithm.

```
C
C Generate an initial ensemble from any QMC walk.
C Move walkers as appropriate for the chosen method. Ap will contain the sum
C of the function A(X) multiplied by the weight, Np is the total weight.
C Begin by assigning each walker to a full circle: Theta = 0, Delta = Tmax.
C (Tmax is the number of angular units in the circle---e.g. 2*Pi or 360.)
C
        Ap = 0
        Np = 0
        Do k=1,Nwalkers
            Theta(k,0) = 0
            Delta(k) = Tmax
        EndDo
        Do i=1,Nsteps
                     .
                     .
                     .
        (Move walkers and branch)
                     .
                     .
                     .
C
C Assign the new ranges. M is the partial sum of the number of offspring in
C the ensemble, which is used to locate the new walkers in the circle.
C
        M = 0
        Do k=1,Nwalker
            N(k) = (number of offspring from walker k)
            Do l=1,N(k)
                Theta(M+1,i) = Theta(k,i-1) + Delta(k)*(l-1)/N(k)
                Delta(M+1) = Delta(k)/N(k)
            EndDo
            M = M + N(k)
        EndDo
        Nwalkers = M
```

```
C
C Once the convergence time Ns has been reached, collect the averages.
C P(k) will be the number of descendents from walker k---hence the weight
C we want.
C
          If ( i > Ns ) Then
             Do k=1,Nwalkers
                P(k) = 0
                Do j=1,Nwalker
                   If (Theta(k,i-Ns) <= Theta(j,i) < Theta(k+1,i-Ns)) Then
                      P(k) = P(k) + 1
                   EndIf
                EndDo
                Ap = Ap + P(k)*A(X(k))
                Np = Np + P(k)
             EndDo
          EndIf
       EndDo
C
C At the end divide Ap by Np to get the pure average.
C
       Apure = Ap / Np
       End
```

7.3.2   Time correlation methods

In Ch. 6 we discussed the time correlation method of obtaining excited-state energies. The method is not restricted to energies, but can be used to determine other properties as well. Consider the time-correlation integrals,[6]

$$H_{IJ}(\tau) = \langle \Psi_I | \mathcal{H} e^{-2\tau\mathcal{H}} | \Psi_J \rangle \tag{7.16}$$

and

$$S_{IJ}(\tau) = \langle \Psi_I | e^{-2\tau\mathcal{H}} | \Psi_J \rangle. \tag{7.17}$$

Although the action of the propagator is to asymptotically project out the ground state, regardless of the starting state Ψ_I, excited state energies may nevertheless be determined from the generalized eigenvalue equation, $\det|\mathbf{H} - E\mathbf{S}| = 0$, as shown in Ch. 6. If we recall that $e^{-\tau\mathcal{H}}$ is the time evolution operator, the QMC state at any time (beginning from the trial function) is $\phi_I(\mathbf{x}, \tau) = e^{-\tau\mathcal{H}} \Psi_I$. Then, for operators such as $\mathcal{A}$ that do not commute with $\mathcal{H}$, one can compute the pure average from

$$A_{IJ}(\tau) \equiv \langle \phi_I(\tau)|\mathcal{A}|\phi_J(\tau)\rangle = \langle\Psi_I|e^{-\tau\mathcal{H}}\mathcal{A}e^{-\tau\mathcal{H}}|\Psi_J\rangle. \tag{7.18}$$

To clarify the operational meaning of Eq. 7.18, we rewrite it by inserting complete sets of (position) states to obtain,

$$\begin{aligned}A_{IJ}(\tau) &= \int \Psi_I(\mathbf{x}_3)\langle\mathbf{x}_3|e^{-\tau\mathcal{H}}|\mathbf{x}_2\rangle\langle\mathbf{x}_2|\mathcal{A}|\mathbf{x}_2\rangle\langle\mathbf{x}_2|e^{-\tau\mathcal{H}}|\mathbf{x}_1\rangle\Psi_J(\mathbf{x}_1)\,d\mathbf{x}_1 d\mathbf{x}_2 d\mathbf{x}_3 \\ &= \int \Psi_I(\mathbf{x}_3)G(\mathbf{x}_3,\mathbf{x}_2;\tau)\mathcal{A}(\mathbf{x}_2)G(\mathbf{x}_2,\mathbf{x}_1;\tau)\Psi_J(\mathbf{x}_1)\,d\mathbf{x}_1 d\mathbf{x}_2 d\mathbf{x}_3. \tag{7.19}\end{aligned}$$

If separate guiding and trial functions are being used, then the diffusion part of the Green's function results in the distribution Ψ_G^2, while the branching term forms the usual pure DMC weights,

$$W(\mathcal{S}(\tau,0)) = \prod_{t=0}^{\tau} e^{-(E_L[\Psi_G(\mathbf{x}(t))]-E_T)\delta t}\,, \tag{7.20}$$

where $\mathcal{S}$ represents the path of the walker from time 0 to τ, and $\mathbf{x}(t)$ the position of the walker at time t. Note that the future-walking weight and the time-correlation weight provide essentially the same quantity, differing only in the method used to obtain them. One method goes to the future, the other goes both to the future and the past. Specifically, by substituting Eq. 7.20 into Eq. 7.19, and interpreting the latter as a path integral, one obtains,

$$\begin{aligned}A_{IJ}(\tau) &= \int_{\mathcal{S}} \Psi_I(\mathbf{x}(\tau))W(\mathcal{S}(\tau,0))A(\mathbf{x}(0))W(\mathcal{S}(0,-\tau))\Psi_J(\mathbf{x}(-\tau))d\mathcal{S} \\ &= \int_{\mathcal{S}} \Psi_I(\mathbf{x}(\tau))A(\mathbf{x}(0))\Psi_J(\mathbf{x}(-\tau))W(\mathcal{S}(\tau,-\tau))d\mathcal{S}. \tag{7.21}\end{aligned}$$

The integral is over all paths, $\mathcal{S}$, of duration 2τ with Ψ_I and Ψ_J sampled at $\pm\tau$ respectively. A is sampled at the midpoint, with the total weight representing the cumulative multiplicity (population) from the initial configuration at time $-\tau$ to the final configuration at time $+\tau$. The contribution to the average is assembled at time τ. This process is shown schematically in Fig. 7.1.

Let us now consider how to use this method to compute pure ground-state averages, i.e. for $\Psi_I = \Psi_J = \Psi_0$. Excited-state, and two-state properties such as transition dipole moments, are discussed in later sections. Although we do not present a formal algorithm here, because it would be essentially the same as that presented in Algorithm 6.1 for excited states, we outline the key points. One begins with a distribution proportional to Ψ^2 from a previous VMC walk. We can take $\Psi_G = \Psi$ because there is only one state to consider. Certain parameters must be chosen. We must pick a maximum time, s_{max}, and m subintervals s_i. These parameters are used to measure the convergence of W to its asymptotic value. For sampling efficiency, each subinterval can be taken to be on the order of the correlation time. The ensemble at the beginning of each subinterval is considered to be the parent of a new walk. We must compute and store $A(\mathbf{X}_k(s_i))$ for each time interval s_i, and collect the cumulative weight $W_k(\mathcal{S}(s_i + s_{max}, s_i - s_{max}))$ just as in Algorithm 6.1. Unlike Algorithm 6.1 however, we must wait until $s_i + s_{max}$ to assemble the averages. The average value of the operator $\mathcal{A}$ is computed as

$$\langle A(s_{max}) \rangle = \frac{\sum_{k,i} W_k(\mathcal{S}(s_i + s_{max}, s_i - s_{max}) A(\mathbf{X}_k(s_i))}{\sum_{k,i} W_k(\mathcal{S}(s_i + s_{max}, s_i - s_{max})}. \tag{7.22}$$

7.3.3 Discussion of Φ_0^2 methods

Let us now compare the above methods and draw from them other, possibly improved, methods. Although we have presented the algorithms separately, there is a direct correspondence between the future-walking method and the time-correlation method. We have already noted that the future-walking weight and the time-correlation weight are both estimates of the quantity Φ_0/Ψ. In future walking, a single weight is com-

Figure 7.1: Schematic diagram of time-correlation sampling method. Each quantity is listed above the time when it is measured. and to the right of the average to which it ultimately contributes.

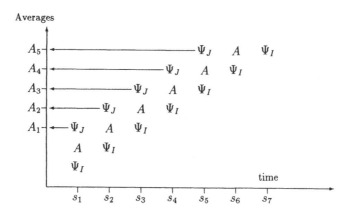

bined with the mixed average to create the pure average. In the time correlation method, on the other hand, two statistically *independent* weights, $W(\mathcal{S}(s_i, s_i - s_{\max}))$ and $W(\mathcal{S}(s_i + s_{\max}, s_i))$ are used to transform Ψ^2 to Φ_0^2. The time-correlation method thus computes two weights, while the future-walking method appears to compute only one. However, the second weight in future walking is implicit in the QMC branching distribution, $f_\infty = \Phi_0 \Psi$. The advantage of the time-correlation form is that the bookkeeping involved is simple: one tracks the weights and averages of a fixed size ensemble; no family trees are required. The advantage of the future-walking method is that the weight of each walker can be kept at unity through branching. Walkers in favorable regions proliferate and those in unfavorable regions expire, as in other branching algorithms. This contrasts with pure DMC in which walkers can have vastly different weights, and ultimately the gap between the largest and smallest weights

Figure 7.2: Schematic of auxiliary walks method.

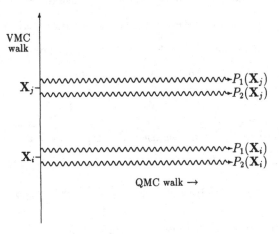

grows exponentially, resulting in a severe loss of precision.

Based on these methods, several variants have been proposed[5-7] that seek to improve them. One of these proceeds by constructing a walk that uses elements of both approaches and, moreover, has the much greater efficiency of VMC, by attaching QMC *auxiliary walks* to a VMC backbone walk.[5, 8] The VMC walk serves as a source of parent configurations; for each parent *two* weights are computed from the asymptotic populations of two branching QMC walks. (These weights are computed as in the future walking algorithm.) Use of auxiliary walks requires additional computation, but depending upon the application this further effort may be offset by the additional stability of the standard branching QMC method. In Fig. 7.2 we show a schematic illustration of this process. The VMC walk is a continual source of configurations for which two independent weights are determined by future walking. The averages are then assembled as in Eq. 7.22.

Another possible modification is to sample $W(\mathcal{S}(s_i + s_{\max}, s_i))$, which is proportional to Φ_0/Ψ, and then *square the weight*[7] to transform Ψ^2 to Φ_0^2. This modification appears to be a very attractive alternative, because it requires much less bookkeeping; but it does *not* produce Φ_0^2! Reference 9 provides a proof of this, and exposes the error in the above logic. Another way to show this is to write,

$$W(\mathcal{S}) = \Phi_0/\Psi + \eta(\mathcal{S}), \tag{7.23}$$

where we have abbreviated a sufficiently long path as $\mathcal{S}$, and η is the random noise or statistical uncertainty in this walk. Integrating over all paths, we obtain,

$$\int_{\mathcal{S}} W(\mathcal{S})d\mathcal{S} = \Phi_0/\Psi + \int_{\mathcal{S}} \eta(\mathcal{S})d\mathcal{S} = \Phi_0/\Psi. \tag{7.24}$$

The noise term vanishes because the average value of η is zero. However, squaring the weight produces,

$$\int_{\mathcal{S}} W^2(\mathcal{S})d\mathcal{S} = {\Phi_0}^2/\Psi^2 + \int_{\mathcal{S}} \eta^2(\mathcal{S})d\mathcal{S}. \tag{7.25}$$

The square of the noise term *does not* vanish. Note, however, that for two statistically independent weights, with independent noise contributions, η_1 and η_2, the final term in Eq. 7.24 does vanish. Squaring the weights does provide an approximation to the pure distribution — but the magnitude of this approximation is not yet fully understood.

7.4   Excited State Properties

We now turn to excited-state properties. Specifically, we will focus on matrix elements between two different electronic states. The simplest method to obtain such a matrix element is to find an approximation similar to Eq. 7.9 for $\tilde{A}_p$. Let us define the following averages:

$$A_v = \langle \Psi_I | A | \Psi_J \rangle, \tag{7.26}$$

$$A_{m_I} = \langle \Phi_I | A | \Psi_J \rangle, \tag{7.27}$$

and,

$$A_{m_J} = \langle \Psi_I | A | \Phi_J \rangle. \tag{7.28}$$

Then a second-order approximation to the matrix element $A_p = \langle \Phi_I | A | \Phi_J \rangle$ is given by

$$\tilde{A}_p = A_{m_I} + A_{m_J} - A_v. \tag{7.29}$$

In Table 7.3 we consider the transition dipole moment between the first two states of the hydrogen atom. We compare analytically computed values of the quantities A_v, A_{m_I}, A_{m_J}, and $\tilde{A}_p$ with the exact value of A_p (all obtained using an approximate trial function). One can see that $\tilde{A}_p$ is a significant improvement over the other estimators. As before, to go beyond the second-order approximation, one can use either the future-walking method or the time-correlation method. As discussed in the treatment of excited states in Ch. 6, since more than one state is involved here, it is important to use a guiding function, Ψ_G, that has non-zero overlap with each state to be considered. A VMC walk can be used to yield the initial distribution Ψ_G^2.

To proceed via future walking, one takes the Ψ_G^2 distribution and conducts auxiliary branching walks for each state of interest. These provide the needed weights. As before, the weights have to be checked for convergence. Further sampling is accomplished by propagating the VMC distribution Ψ_G^2 and repeating the procedure. The results of this method for the Li atom, summarized in Table 7.4, are in very good agreement with experiment, as well as with other high-accuracy *ab initio* methods. Using auxiliary QMC walks allows one to use the fixed-node approximation to sample from the various excited-state distributions. This can be an advantage in cases where transient-estimator methods, such as the time-correlation method, fail to converge rapidly enough to the desired state.

Table 7.3: Comparison of various approximations to the transition dipole moment of H from the $1s$ (state 0) to $2p$ (state 1) states. Orbitals exponents are 0.9 and 0.45, respectively.[8]

Quantity	Analytic value	% Error
A_v	0.8277	11.1
$A_{m_{1s}}$	0.6782	9.0
$A_{m_{2p}}$	0.8981	20.6
$\tilde{A}_p$	0.7485	0.5
A_p	0.7449	0

To proceed using the time-correlation method is a straightforward extension of Algorithm 6.1 and the ground state property discussion of Sec. 7.3.2. To compute A_{IJ} from Eq. 7.21, one needs a walk guided by Ψ_G that overlaps all the states of interest. In addition to the trial functions describing each state, one needs the quantities,

$$F_I(\mathbf{X}) = \frac{\Psi_I(\mathbf{X})}{\Psi_G(\mathbf{X})} \tag{7.30}$$

and the weights W of Eq. 7.20. Just as in Ch. 6, the transient nature of the time correlation method requires that Ψ_I and Ψ_J not be too simple. Choosing Ψ_I and Ψ_J as the eigenfunctions obtained from $\mathbf{H} - E\mathbf{S} = 0$ at time $s_{\max}$ is helpful, as this projects out the ground state and guarantees that the excited states are orthogonal. Properties can then be computed as

$$A_{IJ} = \frac{\sum_{k,i} F_I(\mathbf{X}_k(s_i + s_{\max})) A(\mathbf{X}_k(s_i)) F_J(\mathbf{X}_k(s_i - s_{\max})) W(\mathcal{S}(s_{\max}, -s_{\max}))}{\sqrt{S_{II} S_{JJ}}},$$

$$\tag{7.31}$$

where S_{II} is the normalization of state I, given by

$$S_{II} = \sum_{k,i} F_I(\mathbf{X}_k(s_i + s_{\max})) F_J(\mathbf{X}_k(s_i - s_{\max})) W(\mathcal{S}(s_{max}, -s_{\max})). \tag{7.32}$$

The relative merits of this approach versus future walking are essentially the same here as they were for ground-state properties.

Table 7.4: Comparison of oscillator strengths and excited-state lifetimes (derived from the transition dipole moment) for the $2^2S \rightarrow 2^2P$ transition of Li.[10]

Method	Oscillator Strength (a.u.)	2^2P Lifetime (seconds)
Hartree-Fock	0.768	26.36
Configuration Interaction	0.753	26.89
Hylleraas expansion	0.748	27.07
QMC A_p	0.742(7)	27.4(4)
Experiment	0.742(1)	27.29(4)

7.5 Static and Dynamic QMC Polarizabilities

Besides multipole moments, polarizabilities and hyperpolarizabilities are of interest when studying the response of a molecule to an electric field. The dipole polarizability, for instance, can be obtained as the expectation value of the derivative of the dipole moment with respect to an electric field (either static or time-dependent). We shall treat derivative quantities in the next chapter; however, using pure DMC it is possible to compute these quantities directly without derivatives.

Let us consider the dynamic dipole polarizability. As indicated, it may be expressed as the derivative of the dipole moment with respect to the frequency-dependent electric field $\mathcal{F}_j(\omega)$, i.e.

$$\alpha_{ij}(\omega) = \frac{\partial \mu_i}{\partial \mathcal{F}_j(\omega)}, \qquad (7.33)$$

where i and j are coordinate components x, y or z. However, from second-order perturbation theory one also has an alternative expression, more suitable to us here,

$$\alpha_{ij}(\omega) = 2 \sum_{K=1}^{\infty} \frac{\langle \Phi_0 | \mu_i | \Phi_K \rangle \langle \Phi_K | \mu_j | \Phi_0 \rangle (E_K - E_0)}{(E_K - E_0)^2 - \omega^2}. \qquad (7.34)$$

This expression may be rewritten in a form that can be addressed by pure DMC, i.e.

$$\alpha_{ij}(\omega) = \alpha_{ij}^+(\omega) + \alpha_{ij}^+(-\omega) \qquad (7.35)$$

where

$$\alpha_{ij}^+(\pm\omega) = \sum_{K=1}^{\infty} \frac{\langle\Phi_0|\mu_i|\Phi_K\rangle\langle\Phi_K|\mu_j|\Phi_0\rangle}{(E_K - E_0) \pm \omega}. \tag{7.36}$$

Why is this form more suitable? Within the pure DMC method one can compute the time-correlation function[11] C_{AB} for two quantities A and B from

$$\begin{aligned} C_{AB}(\tau_2 - \tau_1) &\equiv \langle A(\tau_1)B(\tau_2)\rangle \\ &= \frac{\langle\phi_0(\tau_1 - s_{\max})A(\tau_1)W(\mathcal{S}(\tau_2,\tau_1))B(\tau_2)\phi_0(\tau_2 + s_{\max})\rangle}{\langle\phi_0(\tau_1 - s_{\max})W(\mathcal{S}(\tau_2,\tau_1))\phi_0(\tau_2 + s_{\max})\rangle}. \tag{7.37} \end{aligned}$$

We can simplify this expression for C_{AB} by combining the weights with ϕ_0 to produce Φ_0, then inserting a sum over states, and setting $\tau = \tau_2 - \tau_1$, yielding

$$C_{AB}(\tau) = \sum_{K=0}^{\infty} \langle\Phi_0|A|\Phi_K\rangle\langle\Phi_K|B|\Phi_0\rangle \, e^{\tau(E_K - E_0)}. \tag{7.38}$$

To exploit this result, we can take the Laplace transform with respect to the variables τ and ω:

$$\int_0^{\infty} e^{-\tau\omega} \left[C_{AB}(\tau) - \langle A\rangle\langle B\rangle\right] d\tau = \sum_{K=1}^{\infty} \frac{\langle\Phi_0|A|\Phi_K\rangle\langle\Phi_K|B|\Phi_0\rangle}{\omega + E_K - E_0}. \tag{7.39}$$

By setting $A = \mu_i$ and $B = \mu_j$, we obtain the desired quantity $\alpha_{ij}^+(\omega)$. Similar expressions may be derived for the hyperpolarizabilities, from correlation functions involving the functions μ_i, μ_j, μ_k, etc., evaluated at three or more times, and using the appropriate weights. We have listed selected results for the dynamic polarizabilities of He obtained by this method in Table 7.5

Exercises

1. Derive Eqs. 7.7 and 7.8 and use them to yield Eq. 7.9. Derive Eq. 7.29 in a similar manner.

2. Derive Eqs. 7.15 and 7.21 for the pure averages.

Table 7.5: Dynamic dipole polarizabilities[a] $\alpha(\omega)$ of He in a.u.[12]

Frequency ω	QMC	Time dependent Hartree-Fock	Full CI	Estimated exact
0.00	1.38(2)	1.3214	1.3846	1.3834(8)
0.10	1.49(2)	1.3354	1.3885	1.3990(8)
0.20	1.45(2)	1.3797	1.4500	1.4485(8)
0.30	1.55(3)	1.4619	1.5431	1.5412(9)
0.40	1.72(5)	1.5995	1.7009	1.6983(11)
0.50	2.03(11)	1.8327	1.9746	1.9705(15)

(a) $\alpha = (\alpha_{xx} + \alpha_{yy} + \alpha_{zz})/3$.

3. Show that both the future-walking and time-correlation weights provide unbiased estimators for Φ_0/Ψ, but that squaring the weight does not provide an unbiased estimator of $\Phi_0{}^2/\Psi^2$.

4. Implement either the future-walking or time-correlation method. Use it to compute the second moments and transition dipole moment of the lowest two states of the one-dimensional harmonic oscillator. Compute the exact value from the known wave functions.

Suggestions for Further Reading

1. A. C. Hurley, *Electron Correlation in Small Molecules* (Academic, New York, 1976).

2. A. D. Buckingham. "Permanent and Induced Molecular Moments and Long-Range Intermolecular Forces," *Advances in Chemical Physics* **12**, 107-142 (1967).

3. D. M. Ceperley and M. H. Kalos, "Quantum Many-Body Problems," in *Monte Carlo Methods in Statistical Physics*, 183-85 (Springer-Verlag, New York, 1979).

4. S. Zhang and M. H. Kalos, "Bilinear Quantum Monte Carlo: Expectations and Energy Differences," *Journal of Statistical Physics* **70**, 515-33 (1993).

References

1. S. A. Alexander, R. L. Coldwell, G. Aissing and A. J. Thakkar, "Calculation of Atomic and Molecular Properties Using Variational Monte Carlo Methods," *International Journal of Quantum Chemistry* **26**, 213-27 (1992).

2. J. Vrbik, M. F. DePasquale and S. M. Rothstein, "Estimating the Relativistic Energy by Diffusion Quantum Monte Carlo," *Journal of Chemical Physics* **89**, 3784-87 (1988).

3. K. S. Liu, M. H. Kalos and G. V. Chester, "Quantum Hard Spheres in a Channel," *Physical Review A* **10**, 303-8 (1974).

4. P. J. Reynolds, R. N. Barnett, B. L. Hammond, R. M. Grimes and W. A. Lester, Jr., "Quantum Chemistry by Quantum Monte Carlo: Beyond Ground-State Energy Calculations," *International Journal of Quantum Chemistry* **29**, 589-96 (1986).

5. R. N. Barnett, P. J. Reynolds and W. A. Lester, Jr., "Monte Carlo Algorithms for Expectation Values of Coordinate Operators," *Journal of Computational Physics* **96**, 258-76 (1991).

6. D. M. Ceperley and B. Bernu, "The Calculation of Excited State Properties with Quantum Monte Carlo," *Journal of Chemical Physics* **89**, 6316-28 (1988).

7. A. L. L. East, S. M. Rothstein and J. Vrbik, "Sampling the Exact Electron Distribution by Diffusion Quantum Monte Carlo," *Journal of Chemical Physics* **89**, 4880-4 (1988).

8. R. N. Barnett, P. J. Reynolds and W. A. Lester, Jr., "Computation of Transition Dipole Moments by Monte Carlo," *Journal of Chemical Physics* **96**, 2141-54 (1992).

9. P. J. Reynolds, "Does Squaring the Quantum Monte Carlo Weights Give the Exact Quantum Probability Distribution?," *Journal of Chemical Physics* **92**, 2118-19 (1990).

10. R. N. Barnett, P. J. Reynolds and W. A. Lester, Jr., "Monte Carlo Determination of the Oscillator Strength and Excited State Lifetime for the Li 2 $^2S \rightarrow 2\ ^2P$ Transition," *International Journal of Quantum Chemistry* **42**, 837-47 (1992).

11. M. Caffarel and P. Claverie, "Development of a Pure Diffusion Quantum Monte Carlo Method Using a Full Generalized Feynman-Kac Formula. II. Applications to Simple Systems," *Journal of Chemical Physics* **88**, 1100-1109 (1988).

12. M. Caffarel, M. Rérat and C. Pouchan, "Evaluating Dynamic Multipole Polarizabilities and van der Waals Dispersion Coefficients of Two-Electron Systems with Quantum Monte Carlo. A Comparison with Some *Ab Initio* Calculations," *Physical Review A* **47**, 3704-17 (1993).

Chapter 8

Derivatives and Finite Differences

In this chapter we consider the computation of the energy and its derivatives. The use of differences and derivatives greatly expands the scope of physical properties that can be estimated by Monte Carlo, because almost every quantity of physical interest, e.g., dipole moments, polarizabilities, spectroscopic line frequencies, etc., can be formulated as a difference or derivative of the energy. This use of the energy is particularly important for the kind of Monte Carlo applications we have been discussing in this book, namely where the energy is determined exactly from the mixed distribution, but other properties, such as the dipole moment, require the sampling of the pure distribution.

The energy due to a small perturbation $\mathcal{H}'$ to the Hamiltonian, such as that due to an electric field, is sought as the solution of the Schrödinger equation,

$$(\mathcal{H} + \lambda \mathcal{H}')\Phi' = E'\Phi', \tag{8.1}$$

where λ is the strength of the perturbation. The perturbed wave function and energies can be expanded in series in λ,

$$\Phi'(\lambda) = \Phi^{(0)} + \lambda \Phi^{(1)} + \lambda^2 \Phi^{(2)} + \cdots \tag{8.2}$$

and

$$E''(\lambda) = E^{(0)} + \lambda E^{(1)} + \lambda^2 E^{(2)} + \cdots, \tag{8.3}$$

221

where $\Phi^{(n)}$ and $E^{(n)}$ are the nth-order wave function and energy.

Alternatively, we may expand the energy in a Taylor series in the strength of the applied perturbation. For instance, if the perturbation is a uniform electric field of strength $\mathcal{F}_i$ (where $i = 1, 2, 3$, represents the x, y, z directions), the energy for small $\mathcal{F}_i$ is given by

$$E(\mathcal{F}_i) = E(0) + \mathcal{F}_i \left[\frac{\partial E}{\partial \mathcal{F}_i} \right]_{\mathcal{F}_i = 0} + \frac{1}{2} \sum_{j=1}^{3} \mathcal{F}_i \mathcal{F}_j \left[\frac{\partial^2 E}{\partial \mathcal{F}_i \mathcal{F}_j} \right]_{\mathcal{F}_i = 0, \mathcal{F}_j = 0} + \cdots, \qquad (8.4)$$

where the first derivative is the dipole moment, the second derivative is the dipole polarizability tensor, and higher derivatives are dipole hyperpolarizabilities. A comparison of Eqs. 8.3 and 8.4 yields a marked similarity — the energy $E^{(n)}$ is related to a sum of derivatives, and λ is the field strength. Hence, we can obtain the dipole moment by applying a small electric field in each direction (one at a time) and observing the change in the energy. We can view this computation as a finite-field approximation of the perturbed energy, or equivalently as a finite difference calculation used to obtain the derivative. Either way, we have obtained the dipole moment without the need to sample the pure distribution.

Electric fields are not the only perturbations in which we are interested. We also are concerned with displacements of nuclear positions. In the Born-Oppenheimer approximation, the electronic energy is a function of the nuclear framework. For small displacements the energy is given by

$$E(\delta q_i) = E(0) + g_i \delta q_i + \frac{1}{2} \sum_{j=1}^{3N} f_{ij} \delta q_i \delta q_j + \cdots \qquad (8.5)$$

where δq_i is the displacement of the i-th component of the 3-N dimensional nuclear position vector, and g_i and f_{ij} are the energy gradient and force constant matrix (in the harmonic oscillator approximation), respectively. The gradient and force constants are useful in determining the electronic potential energy surface of a molecule.

One may also *combine* electric field and geometry perturbations, to compute quantities such as the intensity of peaks in the infra-red spectrum of a molecule. (The spectrum depends on the change in dipole moment corresponding to a change in geometry.)

For traditional electronic structure methods, the derivatives of the energy can be evaluated either numerically by finite differences, or analytically by differentiating the appropriate energy expressions. This is true in QMC also. In the following we discuss strategies to implement energy derivatives in QMC. Currently, few QMC derivative results are available. If we consider the importance of derivative properties in quantum chemistry, we can expect continued interest in the computation of such properties by QMC.

8.1 Finite Differences and Correlated Sampling

Traditional *ab initio* basis set methods yield approximate values of properties that can be determined to arbitrary precision. In such approaches, the precision of energy differences is primarily governed by the degree to which the wave functions can be converged. It is important to distinguish between *accuracy* and *precision*. Accuracy is determined by the approximation, e.g., Hartree-Fock versus CI. Monte Carlo methods are very *accurate*, but their precision is limited by statistical uncertainty. This limited precision can lead to problems when taking finite differences.

Consider a finite difference calculation of the energy gradient using a standard two-point formula. The derivative can be approximated by.

$$\frac{dE}{dq_i} \approx \frac{E(\delta q_i) - E(0)}{\delta q_i}.$$
(8.6)

If the two energy estimates are statistically uncorrelated, then the statistical error σ_d of the derivative is related to the statistical error σ_e of the energies (assuming σ_e is the same for both energies) by,

$$\sigma_d \propto \sigma_e / \delta q_i. \tag{8.7}$$

Typical values of δq_i in a finite difference method can be quite small, e.g. $\delta q_i = 0.01$ to 0.05 bohr is a common range. This results in a statistical uncertainty in the derivative 20 to 100 (or greater) times that of the uncertainty in the energies.

One approach to this problem is the method of *correlated sampling* (cf. Ch. 2). In this method, a Monte Carlo algorithm is constructed so that the statistical fluctuations in $E(0)$ and $E(\delta q_i)$ are correlated. In the limiting case where the two local energies, denoted as $E_L(\mathbf{x}; 0)$ and $E_L(\mathbf{x}; \delta q_i)$, are related pointwise by a constant offset, correlated sampling of the *difference* would have zero statistical uncertainty. In reality, the variance of the difference increases as the perturbation increases, because the correlation between the perturbed and unperturbed systems decreases. For a finite difference calculation of a derivative, the perturbation is necessarily small, and one expects strong correlation between the local energies — a situation greatly improved over that implied by Eq. 8.7.

Let us first consider correlated sampling in VMC (see also Ch. 2). We use a VMC walk to generate an ensemble of N_c configurations denoted by $\{\mathbf{X}_k^{(i)}\}$, where k is the ensemble index and i is the time index, sampled from a reference function $\Psi(\mathbf{x}; 0)$ taken to be the $\delta q_i = 0$ case. The reference energy is given by the usual Monte Carlo average, i.e.

$$E(0) \approx \frac{1}{MN_c} \sum_{i=1}^{M} \sum_{k=1}^{N_c} E_L(\mathbf{X}_k^{(i)}; 0), \tag{8.8}$$

where the average is over M time steps and N_c configurations. The key to correlated sampling is that the *same* configurations can be used to evaluate the perturbed system. The energy of the perturbed system can be obtained by simply reweighting the energies (computed now with the function $\Psi(\mathbf{x}, \delta q_i)$), namely

$$E(\delta q_i) \approx \frac{\sum_{i=1}^{M} \sum_{k=1}^{N_c} E_L(\mathbf{X}_k^{(i)}; \delta q_i) W(\mathbf{X}_k^{(i)}; \delta q_i)}{\sum_{i=1}^{M} \sum_{k=1}^{N} W(\mathbf{X}_k^{(i)}; \delta q_i)} \tag{8.9}$$

where

$$W(\mathbf{X}_k^{(i)}; \delta q_i) = \frac{\Psi^2(\mathbf{X}_k^{(i)}; \delta q_i)}{\Psi^2(\mathbf{X}_k^{(i)}; 0)}. \tag{8.10}$$

The resulting energy difference $E(\delta q_i) - E(0)$, has a variance that depends on the correlation of the two samples as well as on the sample size, and the correlation can be rather large.. Operationally, one estimates the variance by computing differences over many large blocks of configurations and calculating the variance among these block averages.

Equations 8.9 and 8.10 can be generalized to the full vector $\delta\mathbf{q}$, thereby enabling one to compute all energy differences in a single walk. Moreover, equations similar to Eq. 8.9 and Eq. 8.10 can be written for any measurable property. The only restriction is that the weights must remain close to unity, otherwise the statistical error would grow rapidly. Because of the possibility of introducing discontinuities, particular care must be exercised in computing gradients in cases where the nodes of $\Psi(\mathbf{x}; 0)$ and $\Psi(\mathbf{x}; \delta q_i)$ do not coincide. In VMC, correlated sampling has been used to compute portions of potential energy surfaces, and to optimize variational parameters of a given Ψ.

Correlated sampling in QMC requires an additional contribution to the weight from the Green's function. For our present purposes we discuss only the DMC case. Exact Green's function methods can be derived in an analogous manner. First, recall that a DMC walk with trial function $\Psi(\mathbf{x}; 0)$ will evolve to a distribution $f_\infty(\mathbf{x}; 0) = \Phi_0(\mathbf{x}; 0)\Psi(\mathbf{x}; 0)$, and a walk with trial function $\Psi(\mathbf{x}; \delta q_i)$ will evolve to $f_\infty(\mathbf{x}; \delta q_i) = \Phi_0(\mathbf{x}; \delta q_i)\Psi(\mathbf{x}; \delta q_i)$. To reweight the distribution $f_\infty(\mathbf{x}; 0)$ to create $f_\infty(\mathbf{x}; \delta q_i)$ requires not only the ratio of trial functions, $\Psi(\mathbf{x}; \delta q_i)/\Psi(\mathbf{x}; 0)$, but also the ratio of the QMC solutions, $\Phi_0(\mathbf{x}; \delta q_i)/\Phi_0(\mathbf{x}; 0)$. Although the function Φ_0 is unknown, it was shown in Ch. 7 that the ratio $\Phi_0(\mathbf{x})/\Psi(\mathbf{x})$ is proportional to the asymptotic population of walkers $P(\mathbf{x})$ with ancestor $\mathbf{x}$, i.e.

$$\Phi_0(\mathbf{x})/\Psi(\mathbf{x}) \propto P_\infty(\mathbf{x}) = \lim_{M \to \infty} \prod_{j=0}^{M} G_B(\mathbf{x}(j+1), \mathbf{x}(j); \tau), \qquad (8.11)$$

where G_B is the DMC branching factor.

The perturbed energy therefore may be written as

$$E^{DMC}(\delta q_i) \approx \frac{\sum_{i=1}^{M} \sum_{k=1}^{N} E_L(\mathbf{X}_k^{(i)}; \delta q_i) W(\mathbf{X}_k^{(i)}; \delta q_i)^{DMC}}{\sum_{i=1}^{M} \sum_{k=1}^{N} W(\mathbf{X}_k^{(i)}; \delta q_i)^{DMC}} \qquad (8.12)$$

where

$$W(\mathbf{X}_k^{(i)}; \delta q_i)^{DMC} \equiv \frac{P_\infty(\mathbf{X}_k^{(i)}; \delta q_i) \Psi^2(\mathbf{X}_k^{(i)}; \delta q_i)}{P_\infty(\mathbf{X}_k^{(i)}; 0) \Psi^2(\mathbf{X}_k^{(i)}; 0)}. \qquad (8.13)$$

The challenge of Eq. 8.13 is in the evaluation of $P_\infty(\mathbf{x})$, which must be separately determined with and without the perturbation δq_i. One must create a method both for evaluating $P_\infty(\mathbf{x})$ efficiently and for keeping the branching factors for the two systems from diverging relative to each other. We list representative correlated-sampled finite-difference results in Table 8.1 for the dipole moment (difference of an electric field), geometric gradient (difference of internuclear distances), and equilibrium distance (internuclear distance at which the gradient is zero), and refer the interested reader to the references given in the table for details.

8.2 Virial and Hellmann-Feynman Theorems

Before discussing general derivative formulations, let us discuss two powerful theorems that apply directly to derivatives: the virial theorem and the Hellmann-Feynman theorem. Both theorems provide simple expressions for energy derivatives, though they are strictly applicable only in certain special cases.

Let us define the scalar quantity V to be the expectation value of the potential energy operator $\mathcal{V}$, and T to be the expectation value of the kinetic energy operator $\mathcal{T}$. Naturally, T and V will be exact only if they are evaluated over the pure distribution, $\Phi_0{}^2$. However, averages over the mixed distribution $\Phi_0\Psi$ and the trial distribution Ψ^2 may be combined as in Ch. 7 to provide an estimate of the exact distribution to second order in the difference $\Phi_0 - \Psi$.

8.2.1 The virial theorem

For central forces (such as the Coulomb force), the virial theorem provides the well-known relation

$$2T + V = -\mathbf{q} \cdot \nabla_q E, \tag{8.14}$$

where $\mathbf{q}$ represents the coordinates of the nuclei, and E is the total energy as a function of molecular geometry. Though Eq. 8.14 holds for physical systems, it also holds when T, V, and E are expectation values from a certain class of approximate wave functions. In particular, a wave function that satisfies the Hellmann-Feynman theorem also will satisfy Eq. 8.14 (see Sec. 8.2.2).

For diatomics, Eq. 8.14 reduces to a simple relationship between T, V, and the energy derivative in terms of the internuclear separation, R, namely,

$$-\frac{dE}{dR} = \frac{(2T + V)}{R}. \tag{8.15}$$

Because $E = T + V$, either T or V can be eliminated from Eq. 8.15 in favor of the energy, i.e.,

$$-\frac{dE}{dR} = \frac{(T + E)}{R} = \frac{(2E - V)}{R}. \tag{8.16}$$

For polyatomic molecules, the virial theorem does not provide direct information on any particular derivative, but it is useful insofar as it provides a measure of the equilibrium molecular structure when the right-hand side of Eq. 8.14 vanishes. Once

Table 8.1: Selected correlated-sampling finite-difference results. All values are in atomic units.

Quantity	Method	Value	Experiment	Ref.
LiH dipole moment	DMC	2.27(3)	2.29	2
H_3^+ equilibrium distance	Bessel	1.652(5)	1.6504^a	3
H_2 equilibrium distance	VMC	1.402(1)	1.401	4
BH equilibrium distance	VMC	2.303(10)	2.329	4
Li_2 gradient (R=3.5 bohr)	VMC	$-0.62(4)$	-0.64^a	5

(a) Best variational calculation.

the equilibrium structure is known, the virial theorem can be used to determine second derivatives for diatomics and triatomics. For example, differentiating Eq. 8.14 with respect to q_j and setting $\partial E/\partial q_j$ to zero yields,

$$\frac{\partial V}{\partial q_j} = \sum_i q_i \frac{\partial^2 E}{\partial q_i \partial q_j}. \tag{8.17}$$

For diatomics, Eq. 8.17 reduces to a simple relationship for d^2E/d^2R. For triatomics one obtains a set of three equations for the three second derivatives. Beyond triatomics, the number of second derivatives exceeds the number of Cartesian coordinates, and one cannot solve for the individual second derivatives. Regardless of the number of centers, a major difficulty with Eq. 8.17 is that in order to evaluate $\partial V/\partial q_j$, one must evaluate both derivatives of V and derivatives of the wave function. In the Monte Carlo approach such derivatives are easily obtained for trial functions, but it is far from trivial to find the derivatives of Φ_0 (see Sec. 8.3).

8.2.2　The Hellmann-Feynman theorem

The Hellmann-Feynman theorem has been the subject of much attention in quantum chemistry because it provides a simple and compact expression for the energy gradient. Although the Hellmann-Feynman theorem is a theorem of quantum mechanics, it need not hold for approximate wave functions. Most standard quantum chemical approaches that use atom-centered basis functions *do not* satisfy it. Use of the Hellmann-Feynman formula is therefore only an approximation in these cases, and often results in poor estimates of geometric gradients. Nevertheless, both exact and Hartree-Fock wave functions *do* satisfy the Hellmann-Feynman theorem. Furthermore, it has been shown recently[4] how one can very nearly satisfy the Hellmann-Feynman theorem for standard basis sets.

The Hellmann-Feynman theorem states that the energy gradient is the average of the expectation value of the gradient of the potential, i.e.

$$\nabla_q E_0 = \frac{\int \Phi_0 \nabla_q \mathcal{V} \Phi_0 \, d\mathbf{x}}{\int \Phi_0{}^2 \, d\mathbf{x}}. \tag{8.18}$$

The reason the Hellmann-Feynman theorem fails can be readily discovered from its derivation. Let us start by writing the energy as the expectation value of $\mathcal{H}$ (assuming Φ_0 is normalized) and taking its gradient:

$$
\begin{aligned}
\nabla_q E_0 &= \nabla_q \int \Phi_0 \mathcal{H} \Phi_0 \, d\mathbf{x} \\
&= \int \Phi_0 \left[\nabla_q \mathcal{H} \right] \Phi_0 \, d\mathbf{x} + \int \left[\nabla_q \Phi_0 \right] \mathcal{H} \Phi_0 \, d\mathbf{x} + \int \Phi_0 \mathcal{H} \left[\nabla_q \Phi_0 \, d\mathbf{x} \right]. \tag{8.19}
\end{aligned}
$$

Since $\mathcal{H}$ is a Hermitian operator, the last two terms of Eq. 8.19 are equal. Furthermore, since Φ_0 is an eigenfunction of $\mathcal{H}$, one has that

$$
\begin{aligned}
2 \int \left[\nabla_q \Phi_0 \right] \mathcal{H} \Phi_0 \, d\mathbf{x} &= 2 E_0 \int \left[\nabla_q \Phi_0 \right] \Phi_0 \, d\mathbf{x} \\
&= E_0 \nabla_q \int \Phi_0^2 \, d\mathbf{x} \\
&= 0. \tag{8.20}
\end{aligned}
$$

The final equality results because the integral of Φ_0^2 is unity since Φ_0 is normalized. Equation 8.19 then reduces to Eq. 8.18, because $\mathcal{J}$ does not depend explicitly on nuclear position.

Table 8.2: Comparison of Hellmann-Feynman and analytic derivatives for an SCF wave function of H_2^+ ($R = 1.0$ bohr). Gaussian basis sets are designated by the number of s, p and d ($ns/np/nd$) functions. QMC results are shown for the mixed and pure averages. The energies are given in order to contrast the rates of convergence. All quantities are in atomic units.

Basis set	Energy	$\langle dV/dR \rangle$	$d\langle E \rangle/dR$
(1/0/0)	–0.44087	0.6158	0.5081
(3/1/0)	–0.44549	0.5797	0.5150
(5/2/0)	–0.44954	0.5561	0.5229
(10/2/0)	–0.45059	0.5441	0.5213
(12/4/1)	–0.45167	0.5214	0.5210
QMC mixed	–0.45180(1)	0.571(4)	
QMC pure	–0.45179(1)	0.526(4)	
Exact	–0.45179	0.5207	0.5207

The critical step in this derivation is that Φ_0 is an eigenfunction of $\mathcal{H}$, a step which fails for most computed Ψs. In the Hartree-Fock method, however, the molecular orbitals are eigenfunctions of the Fock operator, and the exact Hartree-Fock solution, i.e., the Hartree-Fock *limit*, can be shown to satisfy the theorem. Unfortunately, when the conditions of the theorem are not satisfied, the Hellmann-Feynman formula is often an unsatisfactory measure of the gradient. This is illustrated in Table 8.2 for H_2^+. We see that the Hellmann-Feynman derivative converges much more slowly with basis set size than the energy and analytic derivative of the energy for typical SCF wave functions — even in this one-electron case! However, we can take advantage of the failure of the Hellmann-Feynman theorem to estimate the quality of molecular trial functions. Specifically, one can make use of the property that the *sum* of the Hellmann-Feynman forces, vanishes for an accurate trial function.

Because of the importance of the Hellmann-Feynman theorem, let us investigate its use in QMC in spite of the above difficulties. We note first an additional difficulty — that in importance-sampled QMC one does not sample the pure estimator $\langle\Phi_0|\nabla_q\mathcal{H}|\Phi_0\rangle$, but rather one samples the mixed estimator, $\langle\Phi_0|\nabla_q\mathcal{H}|\Psi\rangle$. The latter clearly does not satisfy the Hellmann-Feynman theorem. Therefore, unless Ψ is very close to the exact wave function, the analog of Eq. 8.18 will not provide an accurate estimate of the energy gradient. However, either the pure distribution or the second-order approximation given in Ch. 7, may be used to correct for this deficiency. To see how well QMC does with and without such correction, QMC estimates of the Hellmann-Feynman derivative are also listed in Table 8.2. One observes that the pure average is substantially better than the mixed average, and is actually quite close to the exact gradient.

A second difficulty with QMC arises with the variance of the Hellmann-Feynman estimator. Expanding the potential energy gradient in Eq. 8.18 yields

$$\nabla_q\mathcal{V} = \sum_{AB}\frac{Z_AZ_B}{R_{AB}^3}(q_B - q_A) - \sum_{iA}\frac{Z_A}{r_{iA}^3}(q_i - q_A)\,, \qquad (8.21)$$

where subscripts A and B refer to nuclei, i designates electrons, Z is the nuclear charge, R_{AB} is the distance between nuclei A and B, and r_{iA} is the distance between electron i and nucleus A. The first term is the derivative due to nuclear repulsion and the second term is the derivative owing to electron-nucleus attraction. Both are straightforward to evaluate. However, the *variance* of the latter term is infinite. To demonstrate this, let us consider the H atom with the simple wave function $\Psi = e^{-ar}$. The Hellmann-Feynman derivative is trivially zero for an atom. But consider the variance of the Hellmann-Feynman derivative in say the x direction, namely,

$$\sigma^2 = \left\langle\frac{x^2}{r^6}\right\rangle - \left\langle\frac{x}{r^3}\right\rangle^2 = \left\langle\frac{\hat{x}^2}{r^4}\right\rangle - \left\langle\frac{\hat{x}}{r^2}\right\rangle^2\,, \qquad (8.22)$$

where $\hat{x}$ is the angular part of the Cartesian coordinate, i.e. $\hat{x} = x/r = \sin\theta$. Now, writing out the first of these two terms, and converting to polar coordinates (r, θ, φ), we have

$$\left\langle \frac{\hat{x}^2}{r^4} \right\rangle = \int e^{-2ar} \frac{\hat{x}^2}{r^4} dx\,dy\,dz = \int \hat{x}^2\,d\theta d\varphi \int\limits_0^\infty r^2 e^{-2ar} \left(\frac{1}{r^4} \right) dr\ . \qquad (8.23)$$

The angular integral is a constant equal to π^2. More significantly, the radial part is
infinite, and will be so in general. During a QMC simulation this infinite variance
will manifest itself in the form of infrequent but large "spikes" in the value of the
Hellmann-Feynman estimator of the derivative. As a result, no reliable estimate can
be obtained.

A method for suppressing these spikes can be obtained by noting that sufficiently
near the nucleus the charge distribution will be spherical. By partitioning the expectation value of the Hellmann-Feynman derivative into a small spherical region of
radius ε centered on the nucleus, and the rest of space ω, one may write

$$\int \Psi \nabla_q \mathcal{V} \Psi dx = \int\limits_\varepsilon \Psi \nabla_q \mathcal{V} \Psi dx + \int\limits_\omega \Psi \nabla_q \mathcal{V} \Psi dx. \qquad (8.24)$$

The latter integral can be evaluated by QMC, while the former is approximately zero.
The value of ε can be varied, and the results extrapolated to $\varepsilon = 0$ in the same manner
as the time.step extrapolation is done in DMC. This separation removes the singular
behavior of the variance, but the variance still can be large.

8.3   Analytic Energy Derivatives

Analytic energy derivative methods have become ubiquitous in *ab initio* computations.
The reason for this is two-fold. First, as illustrated in Table 8.2, the Hellmann-
Feynman derivative often is less reliable than the analytical derivative. Second, one
can compute accurate second and third derivatives by analytical methods, whereas
finite difference schemes are often costly and inaccurate.

It is worth making a few points about the cost of such derivative calculations in
Monte Carlo. With Monte Carlo, all properties of interest must be sampled during

the walk. Complicated expressions to evaluate can thus greatly slow the process. However, unlike the energy, which must be evaluated at every time step since it is needed to determine the next step, other properties may be sampled at significantly less frequent intervals. Even though this results in fewer sample points, such points are less correlated, and thus more efficiently placed.

8.3.1 Analytic energy gradients

Let us begin with the QMC energy estimator of the ground-state energy,

$$E_0 = \langle E_L \rangle_{f_\infty} = \frac{\int \Phi_0 E_L \Psi \, d\mathbf{x}}{\int \Phi_0 \Psi \, d\mathbf{x}}. \tag{8.25}$$

The gradient of this expression (recalling that one has to take derivatives of Ψ and Φ_0 as well as of E_L in both the numerator and denominator) is

$$\begin{aligned}
\nabla_q E_0 &= \langle \nabla_q E_L \rangle + \left\langle E_L \frac{\nabla_q \Psi}{\Psi} \right\rangle - E_0 \left\langle \frac{\nabla_q \Psi}{\Psi} \right\rangle \\
&+ \left\langle E_L \frac{\nabla_q \Phi_0}{\Phi_0} \right\rangle - E_0 \left\langle \frac{\nabla_q \Phi_0}{\Phi_0} \right\rangle.
\end{aligned} \tag{8.26}$$

(The subscript f_∞ will be assumed for all averages unless otherwise stated.) Except for the terms containing derivatives of Φ_0, there is no difficulty in evaluating Eq. 8.26. The derivatives of Φ_0, while not obvious, can be evaluated using future walking or pure DMC methods. Before doing so, let us derive an approximation for Eq. 8.26 by replacing the derivatives of Φ_0 with those of Ψ to obtain,

$$\nabla_q E_0 \approx \langle \nabla_q E_L \rangle + 2 \left\langle E_L \frac{\nabla_q \Psi}{\Psi} \right\rangle - 2 E_0 \left\langle \frac{\nabla_q \Psi}{\Psi} \right\rangle. \tag{8.27}$$

The derivatives of Ψ are readily obtained from the known analytical form of Ψ. The term $\nabla_q E_L$ also depends only on Ψ. However, this term is complicated because it involves derivatives of the kinetic energy operator, resulting in the need to evaluate the two-electron operator,

$$\sum_q \nabla_q \sum_j \nabla_j^2 \Psi. \tag{8.28}$$

Such operators are cumbersome to evaluate for determinantal wave functions.

It also is important to insure that $\nabla_q E_L$ has a finite variance, or we may have the same problems we encountered earlier with the Hellmann-Feynman theorem derivatives. Here, however, the derivative of the local kinetic energy will cancel the derivative of the potential energy at the origin if the electron-nucleus cusp condition is satisfied. This eliminates the divergence of the variance, though it may still be large. (The method of evaluating the derivative within a radius ε of each nucleus, described in Sec. 8.2.2, is also applicable here.) Even if the singularities of the kinetic and potential energy derivatives do not cancel exactly, one expects that partial cancellation will lead to reduced fluctuations (except at the origin of course) relative to the Hellmann-Feynman derivative, where only the potential term is present. The second and third terms in Eq. 8.27 are explicit corrections to the Hellmann-Feynman formula.

Now let us return to the evaluation of the exact expression for the derivative, Eq. 8.26. Let us begin by rewriting Eq. 8.25 in pure DMC form, i.e.

$$E_0 = \frac{\int \Psi^2(\mathbf{x})W(\mathbf{x})E_L(\mathbf{x})\,d\mathbf{x}}{\int \Psi^2(\mathbf{x})W(\mathbf{x})\,d\mathbf{x}}, \tag{8.29}$$

where the weight W is the cumulative branching defined by

$$W(\mathbf{x}_n) = \prod_{i=1}^{n} e^{-\{1/2[E_L(\mathbf{x}_i)+E_L(\mathbf{x}_{i-1})]-E_T\}\tau}. \tag{8.30}$$

Equation 8.30 was shown in Ch. 7 to be related to the ratio of the exact to the trial wave functions. Using the above two equations, we can now generate an expression for the derivative that depends only on analytically computable quantities, namely

$$\nabla_q E_0 = \langle \nabla_q E_L \rangle \;+\; 2\left\langle E_L \frac{\nabla_q \Psi}{\Psi} \right\rangle - 2E_0 \left\langle \frac{\nabla_q \Psi}{\Psi} \right\rangle$$
$$+ \left\langle E_L \frac{\nabla_q W}{W} \right\rangle - E_0 \left\langle \frac{\nabla_q W}{W} \right\rangle. \tag{8.31}$$

To evaluate Eq. 8.31 one can conduct a non-branching, i.e. pure, DMC walk and accumulate the weight W for several values of the convergence length n. To obtain the derivatives of W, one may use Eq. 8.30. Simplifying the notation by defining $U_i \equiv (1/2)[E_L(\mathbf{x}_i) + E_L(\mathbf{x}_{i-1})] - E_T$, leads to the derivative of the product in Eq. 8.30 as a sum of successively differentiated terms of the form

$$
\begin{aligned}
\frac{\partial W}{\partial q_i} &= \frac{\partial e^{-U_1 \tau}}{\partial q_i} e^{-U_2 \tau} e^{-U_3 \tau} \cdots \\
&\quad + e^{-U_1 \tau} \frac{\partial e^{-U_2 \tau}}{\partial q_i} e^{-U_3 \tau} \cdots + \cdots.
\end{aligned}
$$

Because the derivative of the exponential is the exponential times the derivative of the exponent, the weight can be factored out leaving the following expression for the derivative of the weight,

$$
\frac{\partial W}{\partial q_i} = -W \sum_{j=1}^{n} \tau \frac{\partial U_j}{\partial q_i}. \tag{8.32}
$$

Substituting Eq. 8.32 into Eq. 8.31 yields the energy gradient in terms of the gradients of Ψ and E_L.

Results for LiH and CuH have been obtained[6] using an expression similar to Eq. 8.31, and results for H_2 were obtained[1] from the approximate expression Eq. 8.27. Some of these results are listed in Table 8.3. The calculation of CuH is of particular note because it is the first all-electron VMC computation of a transition metal system.

An additional use of having derivatives of the energy available, is that this information can be used in conjunction with energy values to construct an improved analytical fit to the potential energy curve. Figure 8.1 shows a simple cubic polynomial fit to four QMC energies for H_2. It exhibits spurious oscillatory behavior, being based on only four data points. If the QMC first derivatives are included in the fit, the curve of Fig. 8.2 results. This curve is visually indistinguishable from the exact curve.

Table 8.3: Selected Monte Carlo analytic gradient results. All values are in atomic units.

Quantity	Method	Value	Experiment	Ref.
H_2 equilibrium distance	DMC	1.403(6)	1.401	1
LiH dipole moment	DMC	2.27(1)	2.29	6
CuH equilibrium distance	VMC	2.77(5)	2.764	8
CuH dipole moment	VMC	1.59(4)	$-^a$	8

(a) Not reported.

8.3.2 Higher derivatives and derivative properties

Many chemically interesting properties can be expressed as derivatives of the energy with respect to geometry or external fields or both. Examples are polarizabilities and hyperpolarizabilities, which are second, third and higher derivatives with respect to an external electric field. Although such properties can be evaluated as differences of finite field energies, for higher derivatives such calculations become increasingly costly and unreliable. Clearly, analytical derivative methods would be preferable.

Let us continue this discussion for the general case. Suppose one requires the derivatives of a property associated with the operator $\mathcal{A}$. If we assume that $\mathcal{A}$ does not commute with $\mathcal{H}$, one needs to obtain the pure average,

$$\langle \mathcal{A} \rangle_{\Phi_0^2} = \frac{\int \Phi_0 \mathcal{A} \Phi_0 d\mathbf{x}}{\int \Phi_0^2 d\mathbf{x}}. \tag{8.33}$$

As we saw in Ch. 7, such averages require two independent weights, W_1 and W_2, i.e.,

$$\langle \mathcal{A} \rangle_{\Phi_0^2} = \frac{\int W_1(\mathbf{x}) A(\mathbf{x}) W_2(\mathbf{x}) \Psi^2(\mathbf{x}) d\mathbf{x}}{\int W_1(\mathbf{x}) W_2(\mathbf{x}) \Psi^2(\mathbf{x}) d\mathbf{x}}. \tag{8.34}$$

The two weights may be determined by two separate side walks originating from each point $\mathbf{X}_i$, or, as in the time-correlation method, W_1 is the cumulative branching from time $\tau - n\delta\tau$ to τ, A is computed at τ, and W_2 is the weight from τ to $\tau + n\delta\tau$.

Figure 8.1: Four-point fit of potential curve for H_2.

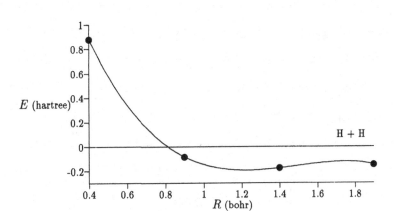

Differentiating Eq. 8.34 with respect to a general parameter λ yields,

$$\frac{\partial}{\partial \lambda} \langle A \rangle_{\Phi_0^2} = \left\langle \frac{\partial A}{\partial \lambda} \right\rangle_{\Phi_0^2} + \left\langle A \left\{ \frac{1}{W_1} \frac{\partial W_1}{\partial \lambda} + \frac{1}{W_2} \frac{\partial W_2}{\partial \lambda} \right\} \right\rangle_{\Phi_0^2} + 2 \left\langle A \frac{1}{\Psi} \frac{\partial \Psi}{\partial \lambda} \right\rangle_{\Phi_0^2}$$

$$- \langle A \rangle_{\Phi_0^2} \left[\left\langle \left\{ \frac{1}{W_1} \frac{\partial W_1}{\partial \lambda} + \frac{1}{W_2} \frac{\partial W_2}{\partial \lambda} \right\} \right\rangle_{\Phi_0^2} + 2 \left\langle \frac{1}{\Psi} \frac{\partial \Psi}{\partial \lambda} \right\rangle_{\Phi_0^2} \right]. \quad (8.35)$$

One may continue this process to obtain second derivatives, such as $\partial^2 E_0 / \partial F \partial q$, and even higher derivatives still.

An approximation to this method has been used in the literature to compute up to fourth derivatives for LiH. The major approximation involved was setting $W_1 W_2$ to W^2. In addition, the expressions were simplified further by using a field-independent trial function, and by assuming the Hellmann-Feynman theorem to hold. Those results are shown in Table 8.4.

Figure 8.2: H$_2$ potential curve fit, including derivatives.

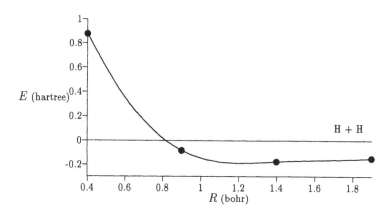

Exercises

1. Derive Eq. 8.26 from Eq. 8.25; then obtain Eq. 8.31 from Eq. 8.26.

2. Show that the derivative of the local kinetic energy cancels the derivative of the potential energy when the trial function satisfies the electron-nucleus cusp condition.

3. Find an explicit expression for the derivative of the kinetic energy of a trial function consisting of a 2×2 determinant (such as for triplet He). What is the expression for a general $N \times N$ determinant?

4. Derive Eq. 8.35 from Eq. 8.34.

Table 8.4: Selected Monte Carlo higher derivatives and derivative properties. All quantities are in atomic units.

Quantity	Method	Value	Experiment	Ref.
LiH polarizability $\alpha_\parallel$	DMC	24.6(4)	26.4[a]	6
LiH hyperpolarizability $\beta_\parallel$	DMC	−440(110)	−686[a]	6
LiH harmonic force constant	DMC	1406(6)	1405.65	7
CuH harmonic force constant	VMC	1814(25)	1941	8

(a) Best variational calculation.

Suggestions for Further Reading

1. D. M. Hirst, *A Computational Approach to Chemistry* (Blackwell Scientific, London, 1990).

2. A. D. Buckingham, "Permanent and Induced Molecular Moments and Long-Range Intermolecular Forces," *Advances in Chemical Physics* **12**, 107-142 (1967).

3. D. S. Chemla and J. Zyss, eds., *Nonlinear Optical Properties of Organic Molecules and Crystals* (Academic, New York, 1987).

4. B. H. Wells, "Green's Function Monte Carlo Methods," in *Methods in Computational Chemistry* **1**, 311-50 (Plenum, New York, 1987).

References

1. P. J. Reynolds, R. N. Barnett, B. L. Hammond, R. M. Grimes and W. A. Lester, Jr., "Quantum Chemistry by Quantum Monte Carlo: Beyond Ground-State Energy Calculations," *International Journal of Quantum Chemistry* **29**, 589-96 (1986).

2. B. H. Wells, "The Differential Green's Function Monte Carlo Method. The Dipole Moment of LiH," *Chemical Physics Letters* **115**, 89-94 (1985).

3. C. A. Traynor and J. B. Anderson, "Parallel Monte Carlo Calculations to Determine Energy Differences Among Similar Molecular Structures," *Chemical Physics Letters* **147**, 389-94 (1988).

4. C. J. Umrigar, "Two Aspects of Quantum Monte Carlo: Determination of Accurate Wavefunctions and Determination of Potential Energy Surfaces of Molecules," *International Journal of Quantum Chemistry: Quantum Chemistry Symposium* **23**, 217-30 (1989).

5. Z. Sun, W. A. Lester, Jr. and B. L. Hammond, "Correlated Sampling of Monte Carlo Derivatives with Iterative-Fixed Sampling," *Journal of Chemical Physics* **97**, 7585-89 (1992).

6. J. Vrbik, D. A. Lagare and S. M. Rothstein, "Infinitesimal Differential Diffusion Quantum Monte Carlo: Diatomic Molecular Properties," *Journal of Chemical Physics* **92**, 1221-27 (1990).

7. J. Vrbik and S. M. Rothstein, "Infinitesimal Differential Diffusion Quantum Monte Carlo Study of Diatomic Vibrational Frequencies," *Journal of Chemical Physics* **96**, 2071-76 (1991).

8. P. Belohorec, S. M. Rothstein and J. Vrbik, "Infinitesimal Differential Diffusion Quantum Monte Carlo Study of CuH Spectroscopic Constants," *Journal of Chemical Physics* **98**, 6401-5 (1993).

Chapter 9

Heavy Atoms

In this chapter we discuss the QMC treatment of heavy-atoms. QMC methods have been applied to systems containing first-row atoms with great success. Atoms further down the periodic table, however, have posed a great computational challenge for these methods. The reason is that a sharp increase in computation time is required to treat atoms of large atomic number Z. The root cause of this increase has been analyzed in a number of ways, but it can be thought of in part as a problem of differing time scales for the valence and core electrons.

A rough estimate of the dependence of the computation time on Z can be obtained as follows. Asymptotically, the time required to compute the energy is dominated by the time needed for evaluation of the Slater determinant and the two-body potential. To update N electrons requires times of order N^3 and N^2 respectively. The evaluation of the two-body potential dominates at small N (because the coefficient of the N^3 term is small), while for large N the evaluation of the Slater determinant dominates. If the energy of a neutral atom of nuclear charge $Z = N$ is sampled M times during a simulation, the computation time T scales for large Z as

$$T \propto MZ^3. \tag{9.1}$$

The value of M is determined by the statistical uncertainty ϵ required to resolve the energy E of interest. Specifically, $\epsilon^2 = \sigma^2/M$, where σ^2 is the variance of the underlying distribution. This variance may be approximated as a sum of a fluctuation term V_1 and a time-step dominated serial correlation term V_2. (We ignore a parallel correlation term due to branching because it can be minimized by avoiding branching.) V_1 is proportional to $E\,\Delta E$, where ΔE may be regarded as the difference between the variational and exact energies.[1] For the important regime of small τ, the dominant contribution to σ^2 is from V_2 and is proportional to $\Delta E/\tau$, where the τ^{-1} dependence results from the reduction in phase space sampled with shorter time steps in the limit of complete serial correlation.[2] If we include both terms we obtain

$$M \propto \frac{E\Delta E + \tau^{-1}\Delta E}{\epsilon^2}. \tag{9.2}$$

For properties of chemical interest which derive from energy differences, e.g. binding energies and ionization potentials, ϵ is essentially constant (typically about 1 kcal/mole), whereas the total energy[3] scales roughly as $E \propto Z^2$. The quantity ΔE appears to be comparable to the correlation energy which scales empirically[4] as $Z^{3/2}$ for $Z < 20$. Finally, to determine the dependence of τ on Z, note that the length scale of the inner electrons decreases as Z^{-1}, e.g. the hydrogenic wave function e^{-Zr} contracts by this factor. Because the distance traveled in one time step is $\Delta R \propto \tau^{1/2}$, $\tau \propto \Delta R^2 \propto Z^{-2}$, so that E and τ^{-1} scale in the same way with Z, i.e. as Z^2. We combine this conclusion with Eqs. 9.1 and 9.2 to find

$$T \propto Z^{6.5}. \tag{9.3}$$

A related argument given elsewhere[2] yielded a *lower* bound of $Z^{5.5}$.

¿From this analysis we see that it is the core electrons that require the smallest time steps and because of their proximity to the nucleus, they contribute most to the total energy. Yet it is the valence electrons that largely determine many chemical properties such as bond strengths, polarizabilities, electron affinities, and ionization potentials, as well as molecular geometries.

Repeating the steps of the above analysis for only the valence electrons, we find that the computation time no longer depends on Z, but on the screened nuclear charge, $Z^{\text{eff}} = Z - N_{\text{core}}$, where N_{core} is the number of electrons in some pre-defined core. Moving across a row of the periodic table, Z^{eff} increases like Z. Thus, by the argument leading to Eq. 9.3, E increases as before and τ decreases. However, the dependence of these quantities on Z^{eff} is considerably weaker than that on Z. In fact, a fit to the SCF valence energies of the second-row elements[5] yields $E \propto (Z^{\text{eff}})^{0.7}$, which is to be compared to the all-electron Z^2 dependence. Because the valence electrons determine the covalent radius, and this scales roughly as $(Z^{\text{eff}})^{-0.35}$ (again based on a fit to second-row atoms), $\tau \propto (Z^{\text{eff}})^{-0.7}$. This result is consistent with the earlier finding that τ^{-1} scales as E. The quantity ΔE should have a somewhat weaker dependence on Z^{eff} than E so that we might assume an upper bound of $\Delta E \propto E \propto (Z^{\text{eff}})^{0.7}$. Finally, in the range of Z^{eff} values, calculation of the two-body potential dominates the Slater matrix inversion, implying $T \propto M(Z^{\text{eff}})^2$. Combining these results gives a computation time-dependence of approximately

$$T \propto (Z^{\text{eff}})^{3.4}. \qquad (9.4)$$

This dependence is a significantly lower power than the previous $Z^{6.5}$. More importantly however, Z^{eff}, unlike Z, remains a small number for *all* atoms.

We thus see that the innermost (core) and outermost (valence) electrons have vastly different time and energy scales. The difficultly posed for QMC is not the inability to sample core and valence electrons, but rather it is the large disparity of time scales. For sufficiently large Z, the Monte Carlo methods discussed thus far will fail to sample completely the different time scales, short of an enormous computational effort that increases with a power of Z. In the following we will outline a number of methods that have been proposed to reduce the power laws. We group these methods into three categories: valence only, in which the core electrons are removed; all-electron approximations, in which the core electrons are retained but are treated approximately; and acceleration, in which all electrons are treated explicitly, but the dynamics of the Monte Carlo simulation is changed.

9.1 Valence Only Methods

¿From the above analysis, it is clear that one approach to reducing the computation time drastically is to eliminate the core electrons, e.g. by use of a pseudopotential. Such methods are long-established in condensed-matter physics and quantum chemistry for computations involving heavy atoms. Two pseudopotential schemes have been most extensively developed in *ab initio* quantum chemistry. The first method, which follows the rigorous pseudopotential formalism of Phillips and Kleinman,[6] is known as the effective-core potential (ECP) method and is the more widely used.[7] The second method[8, 9] is a model potential (MP) approach that is closely related to the "frozen-core" SCF method, an approach that approximates the core orbitals produced in the all-electron SCF method.

The first step in both these methods is to make the frozen-core approximation by approximately partitioning the exact wave function into an antisymmetrized product of core and valence wave functions, i.e.

$$\Phi \approx \mathcal{A}\Psi_{core}\Psi_{val}. \tag{9.5}$$

There is the implicit assumption here that Ψ_{core} does not change with chemical environment. We can then write a valence Schrödinger equation as

$$\mathcal{H}_{val}\Psi_{val} = E_{val}\Psi_{val}, \tag{9.6}$$

where

$$\mathcal{H}_{\text{val}} = \sum_{i=1}^{N_{\text{val}}} \left(-\frac{1}{2} \nabla_i^2 - \sum_\alpha \frac{Z_\alpha^{\text{eff}}}{r_{i\alpha}} + \sum_{j<i}^{N_{\text{val}}} \frac{1}{r_{ij}} \right) + \sum_\alpha \sum_{\beta<\alpha} \frac{Z_\alpha^{\text{eff}} Z_\beta^{\text{eff}}}{R_{\alpha\beta}} + \mathcal{U}. \tag{9.7}$$

The indices i and j refer to the valence electrons while α and β refer to nuclei. The last term, $\mathcal{U}$, is a *non-local* pseudopotential containing terms arising from both core-valence repulsion and antisymmetry. Its form depends on the approach chosen.

9.1.1 Effective core potentials

To construct the pseudopotentials used in ECPs, we begin by writing the Fock equation (see Ch. 5) for a valence atomic orbital in the form,

$$\left[-\frac{1}{2}\nabla^2 - \frac{Z}{r} + \frac{l(l+1)}{2r^2} + \mathcal{V}_{\text{val}} + \mathcal{V}_{\text{core}} \right] \varphi_l = \varepsilon_l \varphi_l. \tag{9.8}$$

The operators $\mathcal{V}_{\text{val}}$ and $\mathcal{V}_{\text{core}}$ contain the Coulomb and exchange potentials of the valence and core electrons respectively. We now eliminate the core electrons by replacing $\mathcal{V}_{\text{val}}$ and $\mathcal{V}_{\text{core}}$ with $\tilde{\mathcal{V}}_{\text{val}}$ and $\mathcal{U}^{ECP}$, and φ_l with pseudo-orbitals χ_l to obtain the valence-only equation,

$$\left[-\frac{1}{2}\nabla^2 - \frac{Z}{r} + \frac{l(l+1)}{2r^2} + \tilde{\mathcal{V}}_{\text{val}} + \mathcal{U}^{ECP} \right] \chi_l = \varepsilon_l \chi_l. \tag{9.9}$$

The pseudo-orbitals are constructed to be smooth and nodeless in the core region, and to match the Hartree-Fock orbitals in the valence region (see Fig. 9.1).

Once the pseudo-orbitals have been constructed from all-electron Hartree-Fock, the potential is found by inverting Eq. 9.9, i.e.

$$\mathcal{U}^{ECP} = \varepsilon_l + \frac{Z}{r} - \frac{l(l+1)}{2r^2} + \frac{[\frac{1}{2}\nabla^2 - \tilde{\mathcal{V}}_{\text{val}}]\chi_l}{\chi_l}. \tag{9.10}$$

Typically the pseudo-orbitals and therefore the pseudopotential are constructed on a grid using the numerical Hartree-Fock procedure. The total pseudopotential is then fit to the analytical form,

$$\mathcal{U}^{ECP} = \sum_{\alpha=1}^{N'_{\text{nuc}}} \sum_{i=1}^{N_{\text{val}}} \left(U_{l_{\text{max}}+1}^\alpha(r_{i\alpha}) + \sum_{l=0}^{l_{\text{max}}} \sum_{m=-l}^{l} |Y_{lm}(\Omega_{i\alpha})\rangle U_l^\alpha(r_{i\alpha})\langle Y_{lm}(\Omega_{i\alpha})| \right), \tag{9.11}$$

Figure 9.1: Typical shapes for the all-electron orbitals φ_l and the pseudo-orbitals χ_l.

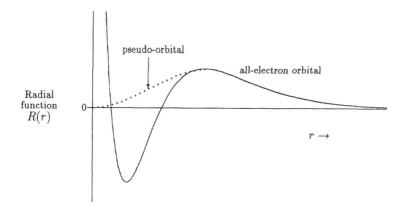

where the sum over α is over only those atoms having a pseudopotential. For atom α, $r_{i\alpha}$ is the distance and $\Omega_{i\alpha}$ is the solid angle of electron i measured from α, $l_{\max}$ is the largest orbital angular momentum among the core orbitals, and U_l^α is a local pseudopotential for atom α that depends only on $r_{i\alpha}$ and the angular momentum l. The spherical harmonics, Y_{lm}, act as projection operators, imposing orthogonality between the missing core and the valence wave function. The functions U_l^α determine the actual form of the ECP and are obtained on a grid. For molecules they are then fit to a linear combination of Gaussians for ease of integration, i.e.

$$U_l(r) = \frac{1}{r^2} \sum_k d_{kl} r^{n_{kl}} e^{-b_{kl}r^2}, \tag{9.12}$$

with n_{kl} taking on values of 0, 1, and 2. In Fig. 9.2 we show plots of some typical U_l's.

The physical interpretation of these potentials is as follows. Each angular momen-

Figure 9.2: Plots of $r^2 U_l$ for Ar.[5]

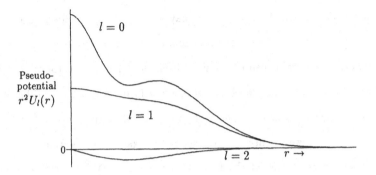

tum component of the valence wave function sees a separate potential. The potential is generally repulsive for angular momenta for which core orbitals exist, e.g. s-orbitals for first-row atoms, as required by the Pauli principle — that is, the existence of a $1s$ orbital in the core requires that the lowest valence s orbital be $2s$-like. However, for angular momenta that do not have a corresponding core orbital, the potential is generally attractive, which represents the de-screening of the nucleus as a valence electron penetrates into the core.

In the usual QMC treatment, the evaluation of the non-local ECP operator poses a problem. One solution is to evaluate it as a local quantity[10-17] by allowing $\mathcal{U}^{ECP}$ to act on Ψ_{val}, a valence importance function, and then dividing this by Ψ_{val}. This evaluation simply adds a term to the local energy, i.e. $U^{ECP}_{local}[\Psi_{\text{val}}] \equiv \Psi^{-1}_{\text{val}} \mathcal{U}^{ECP} \Psi_{\text{val}}$. The first term on the right-hand side of Eq. 9.11 is unaffected by this manipulation because it is already a multiplicative (local) factor. The remaining term becomes,

$$\sum_{\alpha=1}^{N'_{\text{nuc}}} \sum_{i=1}^{N_{\text{val}}} \sum_{l=0}^{l_{\text{max}}} \sum_{m=-l}^{l} Y_{lm}(\Omega_{i\alpha}) U_l^\alpha(r_{i\alpha}) \frac{\langle Y_{lm}(\Omega_{i\alpha})|\Psi_{\text{val}}\rangle}{\Psi_{\text{val}}}. \tag{9.13}$$

Although this localization procedure may seem exact, in analogy to the mixed estimator for E_L, there is a subtle approximation that is being made. In all-electron QMC the exact eigenvalue can be obtained from the mixed estimator of E_L because $\mathcal{H}$ is Hermitian. The present development might indicate that one can still obtain the exact energy eigenvalue of the valence Hamiltonian when $\mathcal{U}^{ECP}$ is part of $\mathcal{H}$. This will rarely occur, however, because one needs the exact eigenfunction to create the correct local representation of the ECP. Otherwise the QMC walk will generate a distribution self-consistent with U_{local}^{ECP} rather than the exact distribution that would be obtained with the correct non-local $\mathcal{U}^{ECP}$. For VMC there is no localization approximation — a random walk simulation using $U_{\text{local}}^{ECP}[\Psi_{\text{val}}]$ will self-consistently produce the distribution Ψ_{val}^2. Recent studies have indicated that high accuracy valence trial functions, either multideterminant[12] or correlated molecular orbital,[18] are required to minimize the localization approximation in QMC.

Evaluation of $\langle \Psi_{\text{val}}|Y_{lm}\rangle$ is itself a rather demanding task, whether accomplished analytically[10] or numerically.[18] In Table 9.1 we show representative results obtained using the ECP method in QMC. These results demonstrate the validity of the use of ECP's in QMC and the degree to which the size of atoms treatable has been extended by the ECP method.

9.1.2 Model potentials

Model potentials provide an alternative approach to the rigorous Phillips-Kleinman scheme which spawned ECP's. The non-local model potential $\mathcal{U}^{MP}$ is chosen to fit some criteria, including coincidence with the Hartree-Fock orbital energy spectrum.

Table 9.1: Selected QMC energies obtained using ECPs.

Quantity	System	QMC	Experiment	Ref.
Ionization Potential	Li	5.412(8)	5.392	10
(eV)		5.345(2)		11
	Na	4.967(4)	5.139	10
	K	4.358(1)	4.341	11
	Be	9.32(1)	9.322	12
	Mg	7.637(26)	7.646	10
		7.657(8)		13
	Al	5.98(3)	5.99	14
	Si	8.24(2)	8.149	21
	Sc	6.54(2)	6.539	21
	Cu	8.8(1)	7.70	21
Electron Affinity	Li	0.611(20)	0.620(7)	10
(eV)		0.557(16)		11
	Na	0.555(21)	0.546(5)	10
	Si	1.42(2)	1.385(5)	21
	K	0.521(11)	0.5012(5)	11
	Cu	1.6(2)	1.22	21
Dissociation Energy	NaH	45.1(1.6)	45.46	10
(kcal/mole)	Na_2	17.2(5)	17.22(2)	10
	Si_2	75.3(1.3)	74.7	21
Excitation Energy	Be $^1S \rightarrow\ ^3P$	2.76(3)	2.73	12
(eV)	Be $^1S \rightarrow\ ^3D$	7.69(3)	7.69	12
	Sc $^2D \rightarrow\ ^4F$	1.5(3)	1.43	15
	Sc$^+$ $^3D \rightarrow\ ^1D$	0.15(2)	0.321	21
	Y $^2D \rightarrow\ ^4F$	1.4(2)	1.36	15

The MP method most widely used[8] in *ab initio* methods is based upon a frozen-core approximation to the Hartree-Fock equations. This method uses a level shift operator together with a model radial potential $\mathcal{V}^{MP}$,

$$\mathcal{U}^{MP} = \sum_c |\varphi_c\rangle B_c \langle \varphi_c| + \mathcal{V}^{MP}. \tag{9.14}$$

The level shift operator, the first term of Eq. 9.14, raises the energy of a core orbital φ_c by the energy B_c. With the appropriate choice of the energy shifts the effect of this operator is to place the core orbitals *above* the valence in energy, so that electrons first fill the valence orbitals. The total energy is, of course, altered (as it also was for ECP's). In form and function, the level shift operator is similar to the angular momentum projection operators in ECP's. The second term in Eq. 9.14, the radial potential, is used to describe the Coulomb repulsion of the core,

$$\mathcal{V}^{MP} = -\frac{Z^{\text{eff}}}{r} U(r), \tag{9.15}$$

where U is typically determined from Hartree-Fock calculations and fit to some simple analytic functional form.

As with ECP's, one can represent this MP locally in QMC as

$$U_{\text{local}}^{MP}[\Psi_{\text{val}}] = \mathcal{V}^{MP} + \sum_c \varphi_c B_c \frac{\langle \varphi_c | \Psi_{\text{val}} \rangle}{\Psi_{\text{val}}}. \tag{9.16}$$

Here the exact solution will result only if the φ_c's are exact. In addition, the local potential will only reproduce the non-local potential if Ψ_{val} is exact.

One feature of the MP method is that the valence orbitals retain the nodal structure of the all-electron valence orbitals, rather than becoming smooth nodeless functions like ECP pseudo-orbitals. Hence one may construct Ψ_{val} such that $\langle \varphi_c | \Psi_{\text{val}} \rangle = 0$ by using the core and valence molecular orbitals obtained from a single Hartree-Fock calculation. Because the Hartree-Fock orbitals are orthogonal from the method, then the projection operators are exactly zero, and only $\mathcal{V}^{MP}$ needs to be evaluated.[19] This is a natural method of localizing the pseudopotential, but it is still an approximation to the correct non-local $\mathcal{U}^{MP}$. As shown in Table 9.2, this method has been used on fewer systems than the ECP approach, but nonetheless has great potential. Its main drawback is that the time step must be considerably smaller than with ECP's because the full nodal structure of the orbitals is maintained.

9.1.3 Pseudo Hamiltonians

In the above methods, the localization of intrinsically non-local pseudopotentials imposes an additional approximation. Localization of the pseudopotential causes the QMC energy to depend *globally* on the quality of Ψ_{val}. This is in contrast to the fixed-node method where the energy depends only on the nodal placement, and the released-node method where the energy does not depend upon the trial function in any way. The problems of using non-local operators in QMC are in fact much deeper. For Hamiltonians containing non-local operators, the exact Green's function need not be positive everywhere — a critical property in QMC. Moreover, in these approaches one gives up the variational principle at the outset.

Instead of standard non-local pseudopotentials consider the general class of pseudo-Hamiltonians (PH's) of the form,

$$\mathcal{H}^{PH} = \mathcal{V} + \nabla \cdot \mathbf{M} \cdot \nabla, \qquad (9.17)$$

where $\mathbf{M}$ is a positive definite "effective mass" tensor. The importance of Eq. 9.17 is that through a proper choice of $\mathbf{M}$, one may reproduce the valence-only electronic wave function obtained with a non-local pseudopotential while keeping both a positive Green's function and the variational principle.

Table 9.2: Selected QMC results using MPs.

Quantity	System	QMC	Experiment	Ref.
Ionization Potential (eV)	Mg	7.571(32)	7.646	19
	Ca	5.878(32)	6.113	19
	Sr	5.573(27)	5.695	19
Electron Affinity (eV)	Cl	3.6(2)	3.615(4)	20

One choice proposed[21] is diagonal in the electrons, i.e.

$$\mathbf{M}_{ij} = -\frac{1}{2}\left(1 + \sum_{\alpha} A_{\alpha}(r_{i\alpha})\right)\delta_{ij}. \tag{9.18}$$

The total electronic Hamiltonian is then given by

$$\mathcal{H}^{PH} = \sum_{i} \nabla_{i}\mathbf{M}_{ii}\nabla_{i} + \mathcal{V}_{ee} + \tilde{\mathcal{V}}_{en} \tag{9.19}$$

where $\mathcal{V}_{ee}$ is the standard electron-electron Coulomb potential, and $\tilde{\mathcal{V}}_{en}$ is the modified electron-nucleus potential,

$$\tilde{\mathcal{V}}_{en} \equiv \sum_{i}\sum_{\alpha} \frac{B_{\alpha}(r_{i\alpha})}{2r_{i\alpha}} + v_{\alpha}(r_{i\alpha}). \tag{9.20}$$

The functions A, B, and v are determined by matching the eigenvectors of $\mathcal{H}^{PH}$ to the valence orbitals of the atom in question. Figure 9.3 shows some typical forms for these functions.

The most obvious way to determine A, B, and v is to match the first three orbitals produced by $\mathcal{H}^{PH}$ with those produced by a non-local ECP. Using the form of Eq. 9.11 for the ECP, with local potentials U_l and one-electron orbitals χ_l, one solves the radial Schrödinger equation,

$$\left(-A\frac{d^2}{dr^2} + (A+B)\frac{l(l+1)}{r^2} - \frac{dA}{dr}\frac{d}{dr}\frac{1}{r} + 2(v-U_l)\right)\chi_l = 0. \tag{9.21}$$

Figure 9.3: Shapes of A, B, and v for Si.[21]

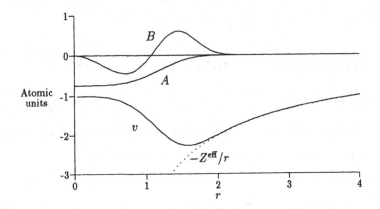

(We refer interested readers to the original paper[21] for details of the method.) The only requirement is that **M** be positive definite, which imposes the conditions

$$A(r) > -1 \tag{9.22}$$

and

$$A(r) + B(r) > -1. \tag{9.23}$$

One result of these conditions is that the lowest one-electron eigenstate of $\mathcal{H}^{PH}$ must be an $l = 0$ state. It is therefore impossible to construct an $\mathcal{H}^{PH}$ that, for example, treats only the $3d$ and $4s$ electrons of a first-row transition metal, since the $3d$ energy lies below the $4s$ in many cases. One can, however, treat a first-row transition metal by including the $3s$, $3p$, $3d$, and $4s$ electrons explicitly. This is not a real limitation of the PH method since it has been shown many times in basis set molecular orbital calculations that the interpenetration of the $3d$ orbitals into the $3s$ and $3p$ orbitals makes it impossible to create an accurate pseudopotential by *any* method in which only the $3d$ and $4s$ orbitals are treated explicitly. In Table 9.3 we have listed results obtained by the PH method. We see that the quality of the results compares favorably with those obtained by the use of ECP's and MP's.

9.2   Approximate All-Electron Methods

Each of the above approaches has been successfully used to take QMC beyond the first-row elements. However we should recall the underlying approximation that the core electrons are inert. Rather than eliminate core electrons, one could instead approximate the core-valence interaction in a manner that alleviates the problem of multiple time scales. We next examine two such approaches: the damped core method, and an effective two-electron pseudopotential method.

9.2.1   All-electron damped-core QMC

The damped-core method attempts to divide the core- and valence-electrons into two separate but linked walks, such that each group can move with its own appropriate time scale[22] (see Fig. 9.4). As before, the wave function is approximately partitioned into core and valence wave functions. The effect of core electrons is simulated by a VMC walk guided by a trial function Ψ_{core} (represented by a few walkers in Fig. 9.4 because of the lack of branching in VMC). The valence electrons (represented by the stack of walkers in Fig. 9.4), which are distinguishable from the core as a result of the factorization of the wave function, are treated by QMC. The valence electrons move in a potential which is due in part to the core electrons. This valence walk is used to obtain the valence energy of the system. Both large-Z effects are addressed: the valence energy is separately evaluated and the valence electrons are separated from the time-scale of the core.

Table 9.3: QMC results obtained using the pseudo-Hamiltonian method.[21]

Quantity	System	QMC	Experiment
Ionization Potential	Na	5.309(3)	5.139
(eV)	Mg	7.753(8)	7.646
	Si	8.19(3)	8.151
	Cl	12.9(1)	12.967
Electron Affinity	Na	0.555(19)	0.546(5)
(eV)	Si	1.39(3)	1.385(5)
	Cl	3.7(1)	3.615(4)
Dissociation Energy	Na_2	19.5(4)	17.22(2)
(kcal/mole)	Mg_2	0.94(44)	1.23
	Si_2	74(2)	74.7
	Cl_2	50(6)	58.0

Core and valence trial functions may be obtained by partitioning the all-electron SCF determinant into its respective orbital groups,

$$\Psi_{\text{core}} = \det|\varphi_1(1) \cdots \varphi_{N_{\text{core}}}(N_{\text{core}})| \tag{9.24}$$

$$\Psi_{\text{val}} = \det|\varphi_{N_{\text{core}}+1}(N_{\text{core}}+1) \cdots \varphi_{N_{\text{val}}+N_{\text{core}}}(N_{\text{val}}+N_{\text{core}})|. \tag{9.25}$$

This is computationally advantageous because the full $N \times N$ determinant need not be evaluated. Rather than MN^3, the computation will scale as $M_{\text{core}}N_{\text{core}}^3 + M_{\text{val}}N_{\text{val}}^3$, where M_{core} and M_{val} are the number of core and valence walkers. (In Fig. 9.4 $M_{\text{core}} = 2$.) The two Hamiltonians are,

$$\mathcal{H}_{\text{core}} = \sum_{s=1}^{N_{\text{core}}} \left(-\frac{1}{2}\nabla_s^2 - \sum_\alpha \frac{Z_\alpha}{r_{s\alpha}} + \sum_{t<s} \frac{1}{r_{st}} \right) + V_{NN} - V_{NN}^{\text{val}} \tag{9.26}$$

and

$$\mathcal{H}_{\text{val}} = \sum_{i=1}^{N_{\text{val}}} \left(-\frac{1}{2}\nabla_i^2 - \sum_\alpha \frac{Z_\alpha}{r_{i\alpha}} + \sum_{j<i}^{N_{\text{val}}} \frac{1}{r_{ij}} + \sum_s^{N_{\text{core}}} \frac{1}{r_{is}} \right) + V_{NN}^{\text{val}}, \tag{9.27}$$

Figure 9.4: Illustration of the damped core method.

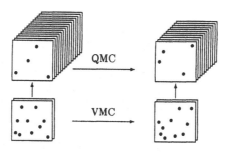

where V_{NN} is the standard nuclear repulsion potential energy, and V_{NN}^{val} is taken to be

$$V_{NN}^{\text{val}} \equiv \sum_{\alpha} \sum_{\beta < \alpha} \frac{Z_{\alpha}^{\text{eff}} Z_{\beta}^{\text{eff}}}{R_{\alpha\beta}}. \tag{9.28}$$

The indices s and t refer to core electrons, i and j to valence electrons, and α and β to nuclei. Note that the valence Hamiltonian is the same as in the pseudopotential case with $\mathcal{U}$ identified as,

$$\mathcal{U} = \sum_{i=1}^{N_{\text{val}}} \left(\sum_{\alpha} \frac{Z_{\alpha}^{\text{eff}} - Z_{\alpha}}{r_{i\alpha}} + \sum_{s}^{N_{\text{core}}} \frac{1}{r_{is}} \right) . \tag{9.29}$$

Because of core-valence factorization, strict orthogonality between the QMC valence solution and VMC core is not maintained, i.e. $\langle \Phi_{val} | \Psi_{\text{core}} \rangle \neq 0$. (This loss of orthogonality is similar to the case of ECP's and MP's where the localization breaks orthogonality to the implied core wave function.) Consequently, at long time the valence QMC solution will be able to occupy core orbitals. To prevent this, one may

Table 9.4: QMC results obtained using the damped-core method.[22]

Quantity	System	QMC	Experiment
Ionization Potential	C	11.2(2)	11.26
(eV)	Si	8.3(2)	8.151
	Ge	8.0(3)	7.899
Electron Affinity	C	1.3(2)	1.268(5)
(eV)	Si	1.3(2)	1.385(5)
	Ge	1.3(3)	1.233(3)

force the valence wave function to remain orthogonal to the core by damping out the branching multiplicity in the core region using a "damped-core" Green's function. (The core region is defined as that region of $3N$-dimensional space where *any* electron approaches the core.) The resulting Green's function is the standard diffusion function with the branching term modulated by a cutoff function $\omega(\mathbf{x})$. Damping may be accomplished using, e.g., a Woods-Saxon type cutoff, i.e.

$$\omega(\mathbf{x}) \equiv \prod_\alpha \prod_i [1 + \exp(-(r_{i\alpha} - \rho_\alpha)/D_\alpha)]^{-1} , \qquad (9.30)$$

where α is the damped-core atom, i loops over electrons, and ρ and D are the cutoff radius and width respectively. When a valence electron approaches an atomic core, $\omega(\mathbf{x})$ approaches zero, and the branching goes to unity. In this limit, one is performing precisely a VMC walk, yielding the variational distribution Ψ_{val}^2 in this core region, whose amplitude is orthogonal to Ψ_{core} by construction. This approximation smoothly connects the fixed-node QMC solution to the variational solution. Numerically, however, there will be a dependence on the choice of ρ and D. Results for some representative systems are presented in Table 9.4.

9.2.2   Effective two-electron potentials

An alternative approach is to focus on reducing the variance of the local energy in the core region by reducing the strength of the two-electron potential.[23] This approach has the advantage that it is local, and that core-valence orthogonality is maintained through the all-electron trial function with the fixed-node approximation. The use of an all-electron trial function, however, has the disadvantages that one must compute the full $N \times N$ determinant, and core and valence electrons still have very different time scales. What is achieved is a core variance reduction.

If we begin with the non-relativistic all-electron Hamiltonian, the two-electron potential $\mathcal{V}_{ee}$ is replaced by,

$$\tilde{\mathcal{V}}_{ee} = \sum_{ij} r_{ij}^{-1} \prod_{\alpha} \left(\omega_{\alpha}(r_{i\alpha}) \omega_{\alpha}(r_{j\alpha}) \right), \tag{9.31}$$

which is a function of not only the interelectronic distance, r_{ij}, but also the electron-nucleus distances $r_{i\alpha}$ and $r_{j\alpha}$. The function ω is again the Woods-Saxon cutoff function, Eq. 9.30. (Note: the cutoff function as defined here is one minus the cutoff function of Ref. 23.) The cutoff function approaches unity as $r \to \infty$ and vanishes as two electrons approach the atomic core.

Having weakened the two-electron Coulomb potential in the core region, the *average* value of the missing part of the atomic two-electron potential must be added back. This average is a function of electron-nucleus distances, so it is most naturally added to $\mathcal{V}_{en}$. Thus,

$$\tilde{\mathcal{V}}_{en}(r) = \begin{cases} U_n(r) - \sum_{\alpha} Z_{\alpha}/r & r > r_t \\ U_{n-1}(r) - \sum_{\alpha} Z_{\alpha}/r & r < r_t \\ , \end{cases} \tag{9.32}$$

where U_n is the angular-momentum averaged potential of the two electrons over an atomic shell,

$$U_n(r) = \frac{1}{n^2} \sum_l (2l+1) u_{n,l}(r), \tag{9.33}$$

and n is the principle quantum number of the atomic shell. The $u_{n,l}$ are obtained from the radial atomic orbitals $\varphi_{n,l}$ by integrating $\tilde{V}_{ee}$ over the coordinates of one of the electrons,

$$u_{n,l}(r_{i\alpha}) = w_\alpha(r_{i\alpha}) \int \sum_{n',l'\neq n,l} \frac{\varphi_{n',l'}(r_{j\alpha})\varphi_{n,l}(r_{j\alpha})\omega_\alpha(r_{j\alpha})dr_{j\alpha}}{r_{ij}}. \tag{9.34}$$

The parameters ρ and D in the cutoff function as well as r_t are chosen to reproduce some known property of the atom. For example, for the Li atom these parameters were obtained by fitting the $^2S \rightarrow\ ^2P$ transition energies, and then the resulting potentials were applied to LiH and Li_2 with the results shown in Table 9.5. The overall efficiency increase, defined as the amount of computational time required to achieve a given precision, was a factor of 5–7 times for these systems. The fact that Li can hardly be considered a heavy atom provides hope for even greater gains for heavier atoms.

Both this and the damped core methods avoid non-local operators and include core-valence correlation approximately. Because the core electrons are represented explicitly, core-valence electronic correlation effects can be taken into account naturally. The price paid for this correlation is the computer time required to maintain a potentially large number of core walkers. In addition, there is an empirical aspect to these approaches, to the extent that they depend upon the choice of parameters.

9.3 Acceleration Methods

All of the above methods concentrate on changing the Hamiltonian of the system to overcome the heavy-atom problem. Alternatively, acceleration methods focus on improving the Monte Carlo simulation *per se*. Because the imaginary time dynamics present in QMC algorithms is not of direct physical interest, one may alter the dynamics in any manner which results in the same steady-state solution. Acceleration

Table 9.5: QMC results obtained using effective two-electron potentials.[23]

Quantity	System	QMC	Experiment
Dissociation Energy	LiH	56.7(6)	57.7
(kcal/mole)	Li_2	25.5(9)	24.7
Excitation Energy (eV)	Li $^2S \to {}^2P$	1.83(1)	1.82

methods hold the promise of an exact stochastic solution to the Schrödinger equation with reduced Z dependence. This is, in principle, preferable to the above approximate methods. However, it is not clear at this time how many powers of Z the acceleration approach can eliminate, nor even how to extend acceleration to QMC — the methods that follow have only been implemented in VMC.

9.3.1   Metropolis acceleration

If one uses a Markov process to generate the VMC distribution, one can change the transition probabilities rather freely as long as detailed balance is maintained (cf. Ch. 1). One way to proceed is with the generalized Metropolis algorithm, which has already been encountered in the VMC discussion of Ch. 2. Here we discuss ways to modify the Metropolis transition probabilities to accelerate VMC.[24]

We begin by reviewing the detailed balance step of the Metropolis algorithm, which requires

$$A(\mathbf{x}, \mathbf{y})G(\mathbf{x}, \mathbf{y})\Psi^2(\mathbf{y}) = A(\mathbf{y}, \mathbf{x})G(\mathbf{y}, \mathbf{x})\Psi^2(\mathbf{x}), \qquad (9.35)$$

where $G(\mathbf{y}, \mathbf{x})$ is the probability of moving from $\mathbf{x}$ to $\mathbf{y}$, Ψ^2 is the desired density function, and A is the acceptance probability given by

$$A(\mathbf{y}, \mathbf{x}) = \min\left(1, \frac{G(\mathbf{x}, \mathbf{y})\Psi^2(\mathbf{y})}{G(\mathbf{y}, \mathbf{x})\Psi^2(\mathbf{x})}\right). \tag{9.36}$$

In the importance-sampled VMC algorithm presented in Ch. 2, G was the Green's function associated with the Fokker-Planck equation for Ψ^2. This choice was predicated on the subsequent use of a Green's function in QMC. But for VMC, any other choice of transition probability is acceptable as long as Eq. 9.35 is satisfied. The simplest choice — which results in "standard" Metropolis sampling — is to take G to be a uniform distribution in a $3N$-dimensional box. However, one also can choose G to reflect the different time scales.

For example, consider the one-electron atom depicted in Fig. 9.5. We may make the box size of the random move from an initial point $\mathbf{x}_1$ to $\mathbf{x}_2$ depend upon $\mathbf{x}_1$. Specifically, let us make it proportional to the electron-nucleus radius, r. This will indeed produce larger moves in the valence region and smaller moves in the core. A modification of such a type has been used to study CuH by VMC.[25] However, this approach has the side effect of producing a time-step bias in the walk unless detailed balance is satisfied. With detailed balance, the move depicted in Fig. 9.5(a) is allowed because there is also a probability for the reverse move. However, in Fig. 9.5(b) the probability of the reverse move is zero! Hence, to maintain detailed balance one must reject the forward move.

One can quickly see that the simple variable-box-size method will be very inefficient. A better way to have valence electrons make larger steps than core electrons[24] is to choose moves in the annular region between $r_1/\delta r$ and $r_1\delta r$, where δr is an adjustable parameter related to the average step size. By choosing the radial displacement in this manner, all moves are reversible. The angular component of the move is then chosen to minimize the serial correlation and to average over the nuclear

Figure 9.5: Illustration of Metropolis acceleration using variable step sizes. Steps are attempted uniformly to a point x_2 within the square having x_1 as its center. The size of the square is chosen to allow walkers to move differently in differing environments. In case (a) the reverse move from x_2 to x_1 lies within the square centered on x_2, and therefore detailed balance can be maintained. In case (b), however, the reverse move is *not* possible. Therefore the indicated forward move must be rejected to maintain detailed balance.

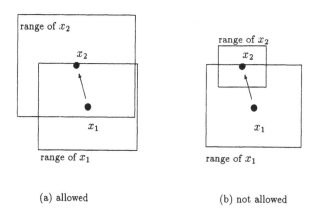

(a) allowed   (b) not allowed

discontinuity as rapidly as possible. Modification must be made, however, to insure detailed balance in a molecule, where r_1 and δr are computed relative to the nearest nucleus. In particular, care must be taken so that no move is allowed in which an electron moves from the sphere of influence of one nucleus to that of another for which the reverse move is impossible. For complete details of this implementation we refer readers to the original paper.[24] Table 9.6 quotes some results of this method, which demonstrate that the algorithm does indeed reduce serial correlation and accelerate

Table 9.6: Comparison of standard and accelerated VMC algorithms.[24]

System	Algorithm	Step^a	Average Acceptance	Decorrelation time
Ne	Standard	0.8	0.769	13.2
	Accelerated	5.0	0.708	2.0
Ar	Standard	0.5	0.808	15.9
	Accelerated	5.0	0.574	2.4

(a) Step refers to radial displacement in accelerated method.

the VMC walk.

Although it is tempting to generalize this method to GFMC methods, unfortunately, this generalization does not seem possible. For example, in DMC the Green's function $G \equiv G_{\text{diff}} G_B$ is derived from the Schrödinger equation. If one changes the dynamics as described above, in effect each component of the walker moves on a different time scale. Although this can be made consistent with a position-dependent G_{diff}, G_B is "global" and it is not clear that a meaningful time step can be defined.

9.3.2 Langevin-based acceleration

A conceptually different method of acceleration[26] is based on the methods of Fourier acceleration that have been used in field theoretic and critical phenomena studies. These methods modify the dynamics in Langevin-based simulations in such a way that the steady-state distribution is unchanged, but is sampled faster, and steps are inherently more decorrelated. One of the important features of this method is that the system jumps more readily among local minima, and that the dynamic critical exponent (i.e., the exponent of the relaxation time) is smaller. The latter is, in effect, the same as reducing the power of Z that scales the computation time. The ability to jump readily among local minima also is important in speeding up sampling and decorrelating moves.

The essence of the idea can be seen as follows. In the VMC algorithm presented in Ch. 2 one has the following Fokker-Planck equation,

$$\frac{\partial f(\mathbf{x}, \tau)}{\partial \tau} = \frac{1}{2} \nabla^2 f(\mathbf{x}, \tau) - \frac{1}{2} \nabla \cdot [\mathbf{F}_Q(\mathbf{x}) f(\mathbf{x}, \tau)]. \tag{9.37}$$

Recall that $\mathbf{F}_Q$ is the drift velocity and $f(\mathbf{x}, \tau)$ is the density function that (here in the absence of branching) converges to $\Psi^2(\mathbf{x})$ at long time. Corresponding to every Fokker-Planck equation is a Langevin equation for the dynamical variables. In this case, that Langevin equation is (in time discretized form)

$$\mathbf{y} = \mathbf{x} + \frac{1}{2} \tau \mathbf{F}_Q(\mathbf{x}) + \xi. \tag{9.38}$$

This is precisely the equation used to update configurations in the Monte Carlo procedure. The quantitiy ξ — the noise term in the Langevin equation — is a vector of Gaussian random numbers (one for each component of $\mathbf{x}$) with zero mean and variance τ. Time-step bias introduced by discretizing the Langevin equation can be removed by the generalized Metropolis acceptance step, in which the transition probability is given by

$$P(\mathbf{y}, \mathbf{x}; \tau) = \min \left\{ 1, \frac{G(\mathbf{x}, \mathbf{y}; \tau) \Psi^2(\mathbf{y})}{G(\mathbf{y}, \mathbf{x}; \tau) \Psi^2(\mathbf{x})} \right\}. \tag{9.39}$$

Perhaps the reader recognizes the above procedure to be precisely the usual VMC algorithm? By seeing its connection to the Fokker-Planck equation, we can take advantage of the work that has been done in field theory which allows one to modify the dynamics of the Fokker-Planck equation and yet obtain the same steady-state solution. The net result is that Eq. 9.37 and Eq. 9.38 can be modified to the following generalized Fokker-Planck equation,

$$\frac{\partial f(\mathbf{x}, \tau)}{\partial t} = \frac{1}{2} \nabla \cdot \mathbf{M} \nabla f(\mathbf{x}, \tau) - \frac{1}{2} \nabla \cdot [\mathbf{M} \mathbf{F}_Q(\mathbf{x}) f(\mathbf{x}, \tau)], \tag{9.40}$$

and associated Langevin equation,

$$\mathbf{y} = \mathbf{x} + \frac{1}{2}\tau\mathbf{M}\mathbf{F}_Q(\mathbf{x}) + \mathbf{M}^{\frac{1}{2}}\xi. \qquad (9.41)$$

The stationary state is the same regardless of the choice of the matrix $\mathbf{M}$, as long as $\mathbf{M}$ is real-symmetric, positive-definite, and independent of coordinates. This means that many different stochastic processes relax to the same equilibrium, but may do so at different rates. By appropriately choosing $\mathbf{M}$, one may accelerate a VMC simulation. (One may of course also slow it down.) The matrix $\mathbf{M}$ mixes coordinates, thereby producing in effect "normal coordinates" that, unlike core and valence coordinates, propagate at comparable speeds.

The remaining problem is how to characterize these modes and to determine $\mathbf{M}$. In an interacting system it may be hopeless to find the "correct" $\mathbf{M}$. However, in field theory as well as in the study of spin glasses, the characterization obtained from noninteracting cases carries over well. Assume that in Eqs. 9.40 and 9.41, $\mathbf{F}_Q$ acts as a force that is derivable from a potential $U = -2\ln|\Psi|$. The electronic configurations seek the (degenerate) minima in this potential, but they are not able to stay precisely at these minima because of the stochastic noise term. If one assumes that near these minima the potential is harmonic, i.e. $U = \frac{1}{2}\sum a_{ij}q_iq_j$, where $\mathbf{q}$ are normal coordinates, it becomes clear that the spread in the eigenvalues of a_{ij} is responsible for the different time scales present. Therefore, a logical choice of $\mathbf{M}$ is

$$M_{ij} = \left\{\frac{\partial^2 U}{\partial x_i \partial x_j}\right\}^{-1}, \qquad (9.42)$$

which is the inverse of the Hessian of U at a minimum. This specification cancels a_{ij}, resulting in only one time scale, and thereby causes all modes to relax at the same rate if U is harmonic. Harmonic potentials are often the non-interacting limits of physical systems. In the case of $U = -2\ln|\Psi|$, however, even a non-interacting Hartree-Fock wave function does not lead to a harmonic U. Moreover, it may be that using M only results in acceleration near the particular minimum that Eq. 9.42 was computed at, and not globally. Nevertheless this VMC acceleration formalism keeps intact the basic structure of the importance-sampled GFMC approach (except for branching). Indeed, Eq. 9.40 appears very similar to the pseudo-Hamiltonian formalism (cf. Eq. 9.17), though the latter uses a diagonal M that *depends* on electronic coordinates. This dependence in the pseudo-Hamiltonian method is introduced because one wants not only to change the dynamics but also the resulting distribution from that of the all-electron system to a valence-only system. On the other hand, a diagonal choice of M in the acceleration approach corresponds to labeling each electron with a separate time scale. This is not totally unreasonable, but less general than desired.

The future of these VMC acceleration methods lies in their application to QMC. However, QMC acceleration will not be feasible until one understands how to change the walker dynamics in a QMC simulation without changing the resulting distribution. Compared to the use of pseudopotentials and other methods discussed in this chapter, acceleration remains a limited method awaiting a breakthrough.

Exercises

1. Use the simple DMC program written for Exercise 6 of Ch. 3 to illustrate the effect of increasing Z on the simulation of two-electron ions. Determine the time step required to achieve a 90% acceptance ratio as Z is varied from one to ten.

2. Show that localizing the non-local ECP, as in Eq. 9.13, is an approximation. What local operator should one sample? How would one sample this operator?

3. One method to choose different time steps for core and valence electrons is to

make the time step proportional to the electron-nucleus distance. Show that such a scheme will violate detailed balance (i.e., moves are possible for which the reverse move is impossible). How does choosing moves between $r_1/\delta r$ and $r_1\delta r$ achieve detailed balance?

Suggestions for Further Reading

1. L. Szasz, *Pseudopotential Theory of Atoms and Molecules* (Wiley, New York, 1985).

2. M. Krauss, W. J. Stevens, "Effective Potentials in Molecular Quantum Chemistry," *Annual Review Physical Chemistry* **35**, 357-85 (1984).

3. W. Müller, J. Flesch and W. Meyer, "Treatment of Intershell Correlation Effects in *Ab Initio* Calculations by Use of Core Polarization Potentials. Method and Application to Alkali and Alkaline Earth Atoms," *Journal of Chemical Physics* **80**, 3297-320 (1984).

4. X.-P. Li, D. M. Ceperley and R. L. Martin, "Cohesive Energy of Silicon by the Green's-Function Monte Carlo Method," *Physical Review B* **44**, 10929-32 (1991).

5. G. G. Batrouni, G. R. Katz, A. S. Kronfeld, G. P. Lepage, B. Svetitsky and K. G. Wilson, "Langevin Simulations of Lattice Field Theories," *Physical Review D* **32**, 2736-47 (1985).

References

1. J. D. Doll, "Monte Carlo Based Electronic Structure Techniques: Analysis and Applications," *Chemical Physics Letters* **81**, 335-38 (1981).

2. D. M. Ceperley, "The Statistical Error of Green's Function Monte Carlo," *Journal of Statistical Physics* **43**, 815-26 (1986).

3. E. Clementi, C. C. J. Roothaan and M. Yoshimine, "Accurate Analytical Self-Consistent Field Functions for Atoms. II. Lowest Configurations of the Neutral First Row Atoms," *Physical Review* **127**, 1618-20 (1962).

4. G. C. Lie and E. Clementi, "Study of the Electronic Structure of Molecules. XXI. Correlation Energy Corrections as a Functional of the Hartree-Fock Density and Its Application to the Hydrides of the Second Row Atoms," *Journal of Chemical Physics* **60**, 1275-87 (1974).

5. W. J. Stevens, H. Basch and M. Krauss, "Compact Effective Core Potentials and Efficient Shared-Exponent Basis Sets for the First- and Second-Row Atoms," *Journal of Chemical Physics* **81**, 6026-33 (1984).

6. J. C. Phillips and L. Kleinman, "New Method for Calculating Wave Functions in Crystals and Molecules," *Physical Review* **116**, 287-94 (1959).

7. L. R. Kahn, P. Baybutt and D. G. Truhlar, "*Ab initio* Effective Core Potentials: Reduction of All-Electron Molecular Structure Calculations to Calculations Involving Only Valence Electrons," *Journal of Chemical Physics* **65**, 3826-53 (1976).

8. V. Bonifacic and S. Huzinaga, "Atomic and Molecular Calculations with the Model Potential Method," *Journal of Chemical Physics* **60**, 2779-86 (1974).

9. S., Huzinaga, L., Seijo, Z., Barandiaran, M. Klobukowski, "The *Ab Initio* Model Potential Method. Main Group Elements," *Journal of Chemical Physics* **86**, 2132-45 (1987).

10. B. L. Hammond, P. J. Reynolds and W. A. Lester, Jr., "Valence Quantum Monte Carlo with *Ab Initio* Effective Core Potentials," *Journal of Chemical Physics* **87**, 1130-36 (1987).

11. M. M. Hurley and P. A. Christiansen, "Relativistic Effective Potentials in Quantum Monte Carlo Calculations," *Journal of Chemical Physics* **86**, 1069-70 (1987).

12. P. A. Christiansen, "Effective Potentials and Multiconfiguration Wave Functions in Quantum Monte Carlo Calculations," *Journal of Chemical Physics* **88**, 4867-70 (1988).

13. P. A. Christiansen and L. A. LaJohn, "Local Potential Error in Quantum Monte Carlo Calculations of the Mg Ionization Potential," *Chemical Physical Letters* **146**, 162-4 (1988).

14. P. A. Christiansen, "Core Polarization In Aluminum," *Journal of Physical Chemistry* **94**, 7865-67 (1990).

15. P. A. Christiansen, "Relativistic Effective Potentials in Transition Metal Quantum Monte Carlo Simulations," *Journal of Chemical Physics* **95**, 361-63 (1991).

16. M. Lao and P. A. Christiansen, "Relativistic Effective Potential Quantum Monte Carlo Simulations for Ne," *Journal of Chemical Physics* **96**, 2161-63 (1992).

17. H.-J. Flad, A. Savin and H. Preuss, "Reduction of the Computational Effort in Quantum Monte Carlo Calculations with Pseudopotentials Through a Change of the Projection Operators," *Journal of Chemical Physics* **97**, 459-63 (1992).

18. L. Mitas, E. L. Shirley and D. M. Ceperley, "Nonlocal Pseudopotentials and Diffusion Monte Carlo," *Journal of Chemical Physics* **95**, 3467-75 (1991).

19. T., Yoshida and K. Iguchi, "Quantum Monte Carlo Method with the Model Potential," *Journal of Chemical Physics* **88**, 1032-34 (1988).

20. T. Yoshida, Y. Mizushima and K. Iguchi, "Electron Affinity of Cl: A Model Potential Quantum Monte Carlo Study," *Journal of Chemical Physics* **89**, 5815-17 (1988).

21. G. B. Bachelet, D. M. Ceperley, M. G. B. Chiocchetti, "Novel Pseudo-Hamiltonian for Quantum Monte Carlo Simulations," *Physical Review Letters* **62**, 2088-91 (1989).

22. B. L. Hammond, P. J. Reynolds and W. A. Lester, Jr. "Damped-Core Quantum Monte Carlo Method: Effective Treatment for Large-Z Systems," *Physical Review Letters* **61**, 2312-15 (1988).

23. J. Carlson, J. W. Moskowitz and K. E. Schmidt, "Model Hamiltonians for Atomic and Molecular Systems," *Journal of Chemical Physics* **90**, 1003-6 (1989).

24. C. J. Umrigar, "Accelerated Metropolis Method," *Physical Review Letters* **71**, 408-11 (1993).

25. P. Belohorec, S. M. Rothstein and J. Vrbik, "Infinitesimal Differential Diffusion Quantum Monte Carlo Study of CuH Spectroscopic Constants," *Journal of Chemical Physics* **98**, 6401-5 (1993).

26. P. J. Reynolds, "Overcoming the Large-Z Problem in Quantum Monte Carlo," *International Journal of Quantum Chemistry* **24**, 679-80 (1990).

Appendix A

Atomic Units

Primary atomic units

Quantity	Unit	SI equivalent
mass	mass of electron, m_e	9.1095×10^{-31} kg
length	Bohr radius, a_0 or bohr	5.29177×10^{-11} m $= 0.529177$Å
angular momentum	$\hbar = h/2\pi$	1.05459×10^{-34} Js
charge	electronic charge, e	1.60219×10^{-19} C
energy	hartree, $\mathcal{E}_a$	4.3598×10^{-18} J

Energy equivalents

1 hartree = 4.3598×10^{-18} J

1.0420×10^{-18} cal

2.626×10^3 kJ mole^{-1}

627.5 kcal mole^{-1}

27.21 eV

2.195×10^5 cm^{-1}

6.580×10^{15} Hz

3.158×10^5 K

Derived atomic units

Quantity	*Unit*	*SI equivalent*
dipole moment	ea_0	8.4784×10^{-30} C m $= 2.54177$ Debye
quadrupole moment	ea_0^2	4.4866×10^{-40} C m^2
electric field	$\mathcal{E}_a e^{-1} a_0^{-1}$	5.1423×10^{11} V m^{-1}
polarizability	$e^2 a_0^2 \mathcal{E}_a^{-1} = a_0^3$	1.6488×10^{-41} C^2 m^2 J^{-1}
first hyperpolarizability	$e^3 a_0^3 \mathcal{E}_a^{-2}$	3.2063×10^{-53} C^3 m^3 J^{-2}
second hyperpolarizability	$e^4 a_0^4 \mathcal{E}_a^{-3}$	6.2360×10^{-65} C^4 m^4 J^{-3}
harmonic force constant	$\mathcal{E}_a a_0^{-2}$	1.5569×10^3 J m^{-2}
cubic force constant	$\mathcal{E}_a a_0^{-3}$	2.9421×10^{13} J m^{-3}
quartic force constant	$\mathcal{E}_a a_0^{-4}$	5.5598×10^{23} J m^{-4}

Appendix B

Evaluating the Trial Function

This appendix is a supplement to Chs. 3 and 5, dealing with the mathematical details of efficiently evaluating the trial function. Specifically, the details here refer to the quantities needed in DMC, as illustrated in Algorithm 3.2, applied to electronic wave functions. In this algorithm each configuration is moved from $\mathbf{x}$ to $\mathbf{y}$ one electron at a time. Each electron is moved by a combination of diffusion and drift, and then accepted or rejected individually. Once all the electrons are moved, branching is carried out, and then the averages are updated. The diffusion-and-drift step requires the generation of a Gaussian random number χ and the evaluation of $\mathbf{F}_Q$, i.e.

$$\mathbf{y} = \mathbf{x} + D\delta\tau\mathbf{F}_Q(\mathbf{x}) + \chi. \tag{B.1}$$

The importance-sampled branching factor $\tilde{G}_B$ requires the evaluation of the local energy,

$$\tilde{G}_B(\mathbf{y}, \mathbf{x}; \delta\tau) = e^{-(\frac{1}{2}[E_L(\mathbf{x}) + E_L(\mathbf{y})] - E_T)\delta\tau}. \tag{B.2}$$

The local energy is also used as an estimator of the exact energy. The acceptance step requires the evaluation of the Metropolis probability,

$$q(\mathbf{y}, \mathbf{x}) = \frac{\Psi^2(\mathbf{y})}{\Psi^2(\mathbf{x})} \frac{\tilde{G}(\mathbf{x}, \mathbf{y}; \delta\tau)}{\tilde{G}(\mathbf{y}, \mathbf{x}; \delta\tau)}, \tag{B.3}$$

which reduces to

$$q(\mathbf{y},\mathbf{x}) = \frac{\Psi^2(\mathbf{y})}{\Psi^2(\mathbf{x})} \exp\{\frac{1}{2}(\mathbf{F}_Q(\mathbf{x}) + \mathbf{F}_Q(\mathbf{y}))[D\delta\tau/2(\mathbf{F}_Q(\mathbf{x}) - \mathbf{F}_Q(\mathbf{y})) - (\mathbf{y} - \mathbf{x})]\},$$

$$(B.4)$$

after substitution of the form of the short-time Green's function. Therefore, at each point $\mathbf{x}$ and $\mathbf{y}$ one needs to evaluate Ψ, $\mathbf{F}_Q$, and E_L. In this discussion we consider trial functions of the form

$$\Psi = \psi_\alpha \psi_\beta S,$$

$$(B.5)$$

where ψ_α and ψ_β are the determinants constructed from the α and β molecular spin orbitals respectively, and the correlation function S is of the Pade-Jastrow form,

$$S = \prod_{u,v} \epsilon^{U_{uv}},$$

$$(B.6)$$

where u and v are combinations of electronic and nuclear positions. The quantum force is related to the gradient of Ψ,

$$\begin{aligned}
\mathbf{F}_Q(\mathbf{x}) &= 2\frac{\nabla\Psi(\mathbf{x})}{\Psi(\mathbf{x})} \\
&= 2\left(\frac{\nabla\psi_\alpha(\mathbf{x})}{\psi_\alpha(\mathbf{x})} + \frac{\nabla\psi_\beta(\mathbf{x})}{\psi_\beta(\mathbf{x})} + \frac{\nabla S(\mathbf{x})}{S(\mathbf{x})}\right).
\end{aligned}$$

$$(B.7)$$

The local energy is a combination of the potential $\mathcal{V}$ (which is independent of Ψ) and the Lapalacian of Ψ, given by

$$E_L(\mathbf{x}) = \mathcal{V}(\mathbf{x}) - 1/2\frac{\sum_{i=1}^{N_{el}} \nabla_i^2 \Psi(\mathbf{x})}{\Psi(\mathbf{x})}.$$

$$(B.8)$$

Here the Laplacian term is

$$\begin{aligned}
\sum_i \frac{\nabla_i^2 \Psi(\mathbf{x})}{\Psi(\mathbf{x})} &= \sum_i \left(\frac{\nabla_i^2 \psi_\alpha(\mathbf{x})}{\psi_\alpha(\mathbf{x})} + \frac{\nabla_i^2 \psi_\beta(\mathbf{x})}{\psi_\beta(\mathbf{x})} + \frac{\nabla_i^2 S(\mathbf{x})}{S(\mathbf{x})}\right. \\
&\quad \left. + \left[\frac{\nabla_i \psi_\alpha(\mathbf{x})}{\psi_\alpha(\mathbf{x})} + \frac{\nabla_i \psi_\beta(\mathbf{x})}{\psi_\beta(\mathbf{x})}\right] \cdot \left[\frac{\nabla_i S(\mathbf{x})}{S(\mathbf{x})}\right]\right).
\end{aligned}$$

$$(B.9)$$

B.1 Efficient Evaluation of the Trial Function

The computation of the potential energy and the correlation function part of the wave function each take on the order of N_{el}^2 operations for a system of N_{el} electrons. However, the computation of the determinant and its derivatives takes order $N_\alpha^3 + N_\beta^3$ operations. One therefore needs to compute the determinant as efficiently as possible. To do so, first notice that in Eqs. B.1 through B.4 only the *ratio* of two determinants is ever required.

Consider the ratio $\Psi(\mathbf{y})/\Psi(\mathbf{x})$ needed in Eq. B.4, ignoring, for now, the correlation functions. Because we are moving a single electron at a time, the two determinants only differ by a single column. The Slater determinant at $\mathbf{x} = (x_1, x_2, \cdots, x_n)$ is

$$\psi(\mathbf{x}) = \det \begin{bmatrix} \varphi_1(x_1) & \varphi_1(x_2) & \cdots & \varphi_1(x_n) \\ \varphi_2(x_1) & \varphi_2(x_2) & \cdots & \varphi_2(x_n) \\ \vdots & \vdots & \ddots & \vdots \\ \varphi_n(x_1) & \varphi_n(x_2) & \cdots & \varphi_n(x_n) \end{bmatrix}, \tag{B.10}$$

which is abbreviated,

$$\psi(\mathbf{x}) = \det|\varphi_1(x_1)\varphi_2(x_2)\cdots\varphi_n(x_n)|. \tag{B.11}$$

We use a lower case ψ to indicate the isolation of the determinant part of the wave function. The spin has been factored out, so n is either N_α or N_β. Moving a single electron i from x_i to y_i (i.e., from $\mathbf{x} = (x_1, \cdots, x_i, \cdots, x_n)$ to $\mathbf{y} = (x_1, \cdots, y_i, \cdots, x_n)$) requires the evaluation of the ratio,

$$\frac{\psi(\mathbf{y})}{\psi(\mathbf{x})} = \frac{\det|\varphi_1(x_1)\cdots\varphi_i(y_i)\cdots\varphi_n(x_n)|}{\det|\varphi_1(x_1)\cdots\varphi_i(x_i)\cdots\varphi_n(x_n)|}. \tag{B.12}$$

To compute this ratio, we first define the Slater *matrix* to be

$$D_{ij}(\mathbf{x}) \equiv \varphi_j(x_i). \tag{B.13}$$

The inverse of the Slater matrix is the adjoint of the matrix of cofactors divided by the determinant of the Slater matrix, i.e.

$$D_{ji}^{-1} = C_{ij}/\psi. \tag{B.14}$$

By definition, the Slater matrix and its inverse must satisfy the relation

$$\sum_{k=1}^{n} D_{ik}(\mathbf{x})D_{kj}^{-1}(\mathbf{x}) = \delta_{ij}. \tag{B.15}$$

The inverse matrix may therefore be used to evaluate the determinant in the usual cofactor expansion. However, the factor of ψ in the denominator of Eq. B.14 makes the inverse matrix most suited to the computation of ratios such as Eq. B.12, i.e.

$$\frac{\psi(\mathbf{y})}{\psi(\mathbf{x})} = \sum_{j=1}^{n} D_{ij}(\mathbf{y})D_{ji}^{-1}(\mathbf{x}). \tag{B.16}$$

The utility of Eq. B.16 is that $\psi(\mathbf{y})$ is never computed. Rather, the inverse matrix at $\mathbf{x}$ is computed, and the desired ratio is obtained from Eq. B.16. The new inverse matrix (at $\mathbf{y}$) is evaluated *only* if the move from $\mathbf{x}$ to $\mathbf{y}$ is *accepted*. In addition, as shown in the next section, it is possible to update the inverse matrix one column at a time, rather than recomputing the entire matrix.

In a similar manner the derivatives are computed as,

$$\frac{\nabla_i \psi(\mathbf{x})}{\psi(\mathbf{x})} = \sum_{j=1}^{n} \nabla\varphi_j(x_i)D_{ji}^{-1}(\mathbf{x}), \tag{B.17}$$

$$\frac{\nabla_i \psi(\mathbf{y})}{\psi(\mathbf{y})} = \frac{\psi(\mathbf{x})}{\psi(\mathbf{y})} \sum_{j=1}^{n} \nabla\varphi_j(y_i)D_{ji}^{-1}(\mathbf{x}), \tag{B.18}$$

$$\frac{\nabla_i^2 \psi(\mathbf{x})}{\psi(\mathbf{x})} = \sum_{j=1}^{n} \nabla^2\varphi_j(x_i)D_{ji}^{-1}(\mathbf{x}), \tag{B.19}$$

and

$$\frac{\nabla_i^2 \psi(\mathbf{y})}{\psi(\mathbf{y})} = \frac{\psi(\mathbf{x})}{\psi(\mathbf{y})} \sum_{j=1}^{n} \nabla^2\varphi_j(y_i)D_{ji}^{-1}(\mathbf{x}). \tag{B.20}$$

Therefore, all the determinant-derived quantities required at each step of the walk can be computed using $D_{ji}^{-1}(\mathbf{x})$.

B.2 Computing the Inverse Slater Matrix

Computation of the inverse Slater matrix can be the single most time consuming step in the QMC walk. Assume that an initial inverse matrix has already been computed for electronic coordinates, $\mathbf{x}$. If only a single electron is moved, only one column of the Slater matrix is changed, yet all the elements of the inverse matrix need to be updated. We have $\mathbf{y}$ equal to $\mathbf{x}$ except that electron i has been moved, as before. Using Eq. B.16, the ratio of the determinant at $\mathbf{y}$ to the determinant at $\mathbf{x}$ is

$$R \equiv \sum_{j=1}^{n} D_{ij}(\mathbf{y}) D_{ji}^{-1}(\mathbf{x}). \tag{B.21}$$

If in the Monte Carlo algorithm this move is rejected, then no further computation is needed. If the move is accepted then the new inverse matrix is related to the old inverse matrix by[1]

$$D_{kj}^{-1}(\mathbf{y}) = D_{kj}^{-1}(\mathbf{x}) - \frac{1}{R} D_{ji}^{-1} \sum_{l=1}^{n} D_{il}(\mathbf{y}) D_{lj}^{-1}(\mathbf{y}) \tag{B.22}$$

for $j \neq i$, and

$$D_{ki}^{-1}(\mathbf{y}) = \frac{1}{R} D_{ki}^{-1}(\mathbf{x}), \tag{B.23}$$

for the elements in column i. One can check this result by substituting Eqs. B.22 and B.23 into Eq. B.15.

To compute the initial inverse matrix one can start with the identity matrix, and replace each column one at a time with the values for each molecular orbital by using Eqs. B.22 and B.23. A brief FORTRAN program to update the inverse matrix follows:

ALGORITHM B.1 Change one column of Slater matrix and find inverse.

```
C=======================================================================
      SUBROUTINE CNGCLM(MATINV,II,NEWCLM,RATIO,TEMP,MATDIM,NWALKER)
C
C Take column II of the Slater determinant and update the inverse.
```

```
C MATINV is the inverse matrix; NEWCLM contains the new column values;
C RATIO is returned with the ratio of the old and new determinants;
C TEMP is a temporary array; MATDIM is the size of the matrix; and
C NWALKER is the number of walkers.
C
C++++++++++++++++++++++++++++++++++++++++++++++++++++++++++++++++++++++++++
C
      REAL*8 MATINV,NEWCLM,RATIO,TEMP
      INTEGER II,MATDIM,NWALKER
C
      DIMENSION MATINV(NI4,MATDIM,MATDIM),NEWCLM(NI4,MATDIM)
      DIMENSION RATIO(NWALKER),TEMP(NCONF)
C
C Begin by computing the ratio of the old and the new determinants.
C
      DO 100 IW = 1,NWALKER
        RATIO(IW) = 0.
  100 CONTINUE
      DO 110 KK = 1,MATDIM
        DO 110 IW = 1,NWALKER
          RATIO(IW) = RATIO(IW) +NEWCLM(IW,KK)*MATINV(IW,II,KK)
  110 CONTINUE
C
C Now update the columns of the inverse matrix, except for column II.
C Loop first over columns.
C
      DO 190 JJ = 1,MATDIM
        IF (JJ .EQ. II) GO TO 190
        DO 140 IW = 1,NWALKER
          TEMP(IW) = 0.
  140   CONTINUE
        DO 150 LL = 1,MATDIM
          DO 150 IW = 1,NWALKER
            TEMP(IW) = TEMP(IW) + NEWCLM(IW,LL)*MATINV(IW,JJ,LL)
  150   CONTINUE
        DO 160 IW = 1,NWALKER
          TEMP(IW) = -TEMP(IW)/RATIO(IW)
  160   CONTINUE
C
C Loop over rows.
C
        DO 180 KK = 1,MATDIM
          DO 180 IW = 1,NWALKER
            MATINV(IW,JJ,KK) = MATINV(IW,JJ,KK) + TEMP(IW)*MATINV(IW,II,KK)
  180   CONTINUE
  190 CONTINUE
C
C Now change column II.
C
      DO 200 KK = 1,MATDIM
        DO 200 IW = 1,NWALKER
          MATINV(IW,II,KK) = MATINV(IW,II,KK)/RATIO(IW)
  200 CONTINUE
      RETURN
      END
```

B.3 Molecular Orbitals and Correlation Functions

The only remaining tasks are to compute the molecular orbitals (MOs) and the correlation functions. We list these here for completeness. The MOs are given as linear combinations of atomic orbitals (AOs), χ. These are either STOs or GTOs.

For Slater-type orbitals (STO's) the following relationships hold

$$\chi_S = x^a y^b z^c r^n e^{-\zeta r}, \tag{B.24}$$

$$\chi_S^{-1} \frac{\partial \chi_S}{\partial x} = \frac{a}{x} - \frac{\zeta x}{r} + \frac{nx}{r^2}, \tag{B.25}$$

and

$$\chi_S^{-1} \nabla^2 \chi_S = \frac{a(a-1)}{x^2} + \frac{b(b-1)}{y^2} \frac{+c(c-1)}{z^2}$$
$$+ \zeta^2 - \frac{2\zeta l}{r} + \frac{2n(l - n/2 - \frac{1}{2})}{r^2}, \tag{B.26}$$

where $l = a + b + c + n + 1$.

For Cartesian Gaussian-type orbitals (GTO's), the analogous equations are

$$\chi_G = x^a y^b z^c \exp(-\zeta r^2), \tag{B.27}$$

$$\chi_G^{-1} \frac{\partial}{\partial x} \chi_G = \frac{a}{x} - 2\zeta x, \tag{B.28}$$

and

$$\chi_G^{-1} \nabla^2 \chi_G = \{a(a-1)x^{a-2} + b(b-1)y^{b-2} + c(c-1)z^{c-2}$$
$$-2\zeta[(2a+1)x^a + (2b+1)y^b + (2c+1)z^c]$$
$$+4\zeta^2(x^{a+2} + y^{b+2} + z^{c+2})\} x^{-a} y^{-b} z^{-c}. \tag{B.29}$$

In both Eqs. B.24 and B.27, the normalization has been omitted. The normalization constant for an STO is

$$N_{STO}(a, b, c, \zeta) = \frac{(2l)!}{(2\zeta)^{2l+1}} \frac{4\pi(2a-1)!!(2b-1)!!(2c-1)!!}{(2a+2b+2c+1)!!}, \tag{B.30}$$

where the function $n!! \equiv 1 \cdot 3 \cdot 5 \cdots n$ $(0!! \equiv 1)$. For a GTO the normalization constant is most easily found by factoring the GTO into x, y, and z components. For example, the x-component is $x^a \exp(-\zeta x^2)$. The normalization for this component is

$$N_{GTO}(a, \zeta) = (\frac{2\zeta}{\pi})^{\frac{1}{4}}(4\zeta)^{\frac{a}{2}}[(2a-1)!!]^{-\frac{1}{2}}. \tag{B.31}$$

The total normalization is given by the product of $N_G(a, \zeta)$, $N_G(b, \zeta)$, and $N_G(c, \zeta)$.

Lastly, we give the necessary formulas for dealing with the linear electron-electron Padé-Jastrow function

$$S = \exp\left\{\sum_{i>j} \frac{ar_{ij}}{1 + br_{ij}}\right\}. \tag{B.32}$$

The gradient of S is

$$S^{-1}\nabla S = \sum_{i>j} \frac{\mathbf{x}}{r} \frac{a}{(1 + br_{ij})^2}, \tag{B.33}$$

while the Laplacian is

$$S^{-1}\nabla^2 S = \sum_{i>j} \frac{a(2r_{ij}^{-1}(1 + br_{ij}) + a)}{(1 + br_{ij})^4}. \tag{B.34}$$

References

1. D. M. Ceperley, G. V. Chester and M. H. Kalos, "Monte Carlo Simulation of a Many-Fermion System," *Physical Review B* **16**, 3081-99 (1977).

Appendix C

Sample Diffusion Monte Carlo Program

In this appendix we give a listing of a FORTRAN77 implementation of the diffusion quantum Monte Carlo algorithm. No importance sampling is used here, and only one-dimensional potentials are implemented.

The most significant parameters are as follows:

- MWALK and MBIN are the dimensions of the walker and histogram arrays. AMASS is the mass of the particle. The value exp(OVERFLO) is the largest number on the computer. TWOPI is 2π.

- NORMWLK is the initial size of the ensemble. Whenever the ensemble grows to MAXWLK, walkers are deleted from the end of the list to return the size to NORMWLK. Whenever the ensemble shrinks below MINWLK, the size is reset to MINWLK.

- NSTEP is the number of steps per block; NBLOCK is the number of blocks. NEQUIL is the number of blocks allowed for the walk to equilibrate.

- TAU is the time step; ET is the initial trial energy.

- The binning of walkers for the histogram of the wave function is controlled by NBIN (number of bins), XINIT (the smallest value of X) and XLAST (the largest value of X).

This program was designed for vector supercomputer architectures. The innermost loop, wherever possible, is over the walkers. This arrangement complicates the structure of the program somewhat, requiring a number of temporary arrays and additional DO loops. However, given that the number of walkers is typically large, this structure guarantees long vector lengths, and thereby an efficient vector code.

A more complete QMC molecular electronic structure program, QuantumMagiC, developed at the University of California at Berkeley, may be obtained from the Quantum Chemistry Program Exchange.

```
      PROGRAM QMONTE
      IMPLICIT REAL*8 (A-H,O-Z)
C
C Set parameters and array sizes.
C
      PARAMETER (MWALK=2000, MBIN=20)
      PARAMETER (AMASS=1.0)
      PARAMETER (OVERFLO=70.0, TWOPI=6.2831853)
      DIMENSION X(MWALK),VOLD(MWALK),V(MWALK),IMULT(MWALK)
      DIMENSION IBINPOP(MBIN)
C
C Here is the input to the program.
C
      DATA MINWLK/250/,MAXWLK/2000/,NORMWLK/1000/,NBLOCK/100/
      DATA NSTEP/100/,TAU/0.01/,ET/1.0/,NEQUIL/20/
      DATA NBIN/15/,XINIT/-4.0/,XLAST/4.0/
C
C Start the random number generator.
C
      ISEED = 18293753
      DUM = URAN(ISEED)
C
C Begin run.
C
      WRITE(*,1000)
 1000 FORMAT(' BLOCK',3X,'EBLOCK',10X,'EGROWTH',10X,'ETRIAL ',
     .10X,'POPLTN')
      D = 1.E0/(2.E0 * AMASS)
      TAUINV = 1.E0/TAU
      NWALKER = NORMWLK
      EAVE = 0.E0
      EAV2 = 0.E0
C
C Initialize walker positions.
```

```
C
      XWIDTH = XLAST - XINIT
      XBINWIDTH = XWIDTH/NBIN
      DO 100 IW=1,NWALKER
        X(IW) = URAN(0)*XWIDTH + XINIT
  100 CONTINUE
C
C Load potential array. POTENT1 is a harmonic oscillator,
C POTENT2 is a Morse oscillator.
C
      CALL POTENT1(MWALK,NWALKER,AMASS,X,V)
C     CALL POTENT2(MWALK,NWALKER,AMASS,X,V)
C
C Perform walk and accumulate statistics.
C
      NBDONE = 0
      DO 200 IBLOCK=1,NBLOCK
        IF (IBLOCK.GT.NEQUIL) NBDONE = NBDONE + 1
        EBLOCK = 0.
        EG = 0.
        DO 140 IT=1,NSTEP
          NOLD = NWALKER
          DO 120 IW=1,NWALKER
            VOLD(IW) = V(IW)
  120     CONTINUE
          CALL AWALK(MWALK,NWALKER,TWOPI,D,TAU,X)
          CALL POTENT1(MWALK,NWALKER,AMASS,X,V)
C         CALL POTENT2(MWALK,NWALKER,AMASS,X,V)
          CALL BRANCH(MWALK,NWALKER,NORMWLK,MAXWLK,MINWLK,
     &                OVERFLO,TAU,ET,IMULT,X,VOLD,V)
          DUM = FLOAT(NOLD)/FLOAT(NWALKER)
          EG = EG + (TAUINV * LOG(DUM) + ET)
          DO 130 IW=1,NWALKER
            EBLOCK = EBLOCK + V(IW)
  130     CONTINUE
  140   CONTINUE
C
C Compute the average value of the potential, EBLOCK, and the
C growth energy, EG, which are both estimates of the total energy.
C The growth energy is used to update the trial energy.
C
        EBLOCK = EBLOCK / NSTEP / NWALKER
        EG = EG/NSTEP
        IF (IBLOCK.LE.NEQUIL) THEN
          ET = 0.5 * ET + 0.5 * EG
        ELSE
          ET = ET + 0.5 * (EG - ET)/NBDONE
        ENDIF
        IF (IBLOCK.GT.NEQUIL) THEN
          EAVE = EAVE + EBLOCK
          EAV2 = EAV2 + EBLOCK**2
          AMIN = XINIT
          DO 160 IBIN=1,NBIN
            AMAX = AMIN + XBINWIDTH
            DO 150 IW=1,NWALKER
              IF(X(IW).GE.AMIN.AND.X(IW).LT.AMAX)
     &          IBINPOP(IBIN) = IBINPOP(IBIN) + 1
  150       CONTINUE
            AMIN = AMAX
  160     CONTINUE
        ENDIF
```

```
C
C Write out results for this block.
C
        WRITE(*,1020) IBLOCK,EBLOCK,EG,ET,NWALKER
 1020   FORMAT(I3,3E17.5,I10)
        IF (NWALKER.GT.NORMWLK) NWALKER = NORMWLK
  200 CONTINUE
C
C Write out grand averages.
C
      EAVE = EAVE/NBDONE
      EAV2 = EAV2/NBDONE
      VAR = SQRT(ABS(EAVE**2 - EAV2)/NBDONE)
      WRITE(*,1030) EAVE,VAR
 1030 FORMAT(/,' ======= GRAND AVERAGES ======= ',/,
     & ' GROUND STATE ENERGY',T37,F12.6,1X,
     &          '+/-',1X,F9.6)
      WRITE(*,1040) (IBINPOP(IBIN),IBIN=1,NBIN)
 1040 FORMAT(/,'PROBABILITY DENSITY HISTOGRAM',
     &        /,15(I5,1X))
C
      STOP
      END
C=====================================================================
      SUBROUTINE AWALK(MWALK,NWALKER,TWOPI,D,TAU,X)
      IMPLICIT REAL*8 (A-H,O-Z)
      DIMENSION X(MWALK)
C
C Move each walker. Use the Box-Muller method to generate random numbers
C with a spherical Gaussian distribution. D is the diffusion constant.
C Gaussian random numbers are generated two at a time; if there are an
C odd number of walkers, generate one extra.
C
      DO 100 IW=1,MIN(NWALKER+1,MWALK),2
        RAND1 = URAN(0)
        AMP = SQRT (-4.*D*TAU*LOG(RAND1))
        THETA = TWOPI * URAN(0)
        DX1 = AMP * SIN(THETA)
        DX2 = AMP * COS(THETA)
        X(IW  ) = X(IW  ) + DX1
        X(IW+1) = X(IW+1) + DX2
  100 CONTINUE
      RETURN
      END
C=====================================================================
      SUBROUTINE BRANCH(MWALK,NWALKER,NORMWLK,MAXWLK,MINWLK,
     &                  OVERFLO,TAU,ET,IMULT,X,VOLD,V)
      IMPLICIT REAL*8 (A-H,O-Z)
      DIMENSION X(MWALK),V(MWALK),VOLD(MWALK),IMULT(MWALK)
C
C Perform the branching. All the walkers are kept on the list X. A
C walker is deleted by replacing it by the last one on the list, and
C reducing the number of walkers by one. Replicas of walkers are
C added to the end of the list.
C
C First compute the branching weight in diffusion Monte Carlo.
C
      DO 110 IW=1,NWALKER
        AMULT = -TAU*(0.5E0*(VOLD(IW) + V(IW)) - ET)
        IMULT(IW) = INT(EXP(AMULT) + URAN(0))
  110 CONTINUE
```

```
      K = 0
  120 K = K + 1
      IF(K.GT.NWALKER) GO TO 150
  130 IF(IMULT(K).GT.0) GO TO 120
C
C Delete dead walkers.
C
      IF(K.EQ.NWALKER) GO TO 140
      IMULT(K) = IMULT(NWALKER)
      V(K) = V(NWALKER)
      X(K) = X(NWALKER)
      NWALKER = NWALKER - 1
      GO TO 130
  140 NWALKER = NWALKER - 1
  150 CONTINUE
      NEWPOP = 0
      DO 160 IW=1,NWALKER
         NEWPOP = NEWPOP + IMULT(IW)
  160 CONTINUE
C
C Replicate and assign X() and V() to newly created walkers.
C
      K = NWALKER
      IW = 0
  170 IW = IW + 1
      IF (IW.GT.NWALKER) GO TO 190
      JJ = 0
  180 JJ = JJ + 1
      IF(JJ.GT.IMULT(IW)-1) GO TO 170
      K = K + 1
C
C Take care of overflowing the ensemble array.
C
      IF(K.GT.MAXWLK) THEN
         IFLAGOV = 1
         NWALKER = NORMWLK
         WRITE(*,*) '    ENSEMBLE OVERFLOW: RETURNING TO NORMWLK'
         GO TO 200
      ENDIF
      V(K) = V(IW)
      X(K) = X(IW)
      GO TO 180
  190 CONTINUE
      NWALKER = NEWPOP
C
C Take care of underflows.
C
      IF(NWALKER.LT.1) THEN
         WRITE(*,*) '    ENSEMBLE UNDERFLOW: RETURNING TO MINWLK'
         NWALKER = MINWLK
      ENDIF
  200 CONTINUE
      RETURN
      END
C======================================================================
      FUNCTION URAN(I)
      IMPLICIT REAL*8 (A-H,O-Z)
C
C Return a uniformly distributed pseudo-random number between 0 and 1.
C If I=0 return next number in sequence.  If I<0 set I to random
C number seed ISEED. If I>0 return ISEED. This is a linear congruential
```

```
C generator that has been coded specifically to avoid integer overflows.
C
      PARAMETER (IA=16807, IM=2147483647, AM=1./IM, IQ=127773, IR=2836)
      SAVE ISEED
C
      IF (I.EQ.0) THEN
        IK1 = ISEED/IQ
        ISEED = IA*(ISEED - IK1*IQ) - IR*IK1
        IF (ISEED.LT.0) ISEED = ISEED + IM
        URAN = AM*ISEED
      ELSE IF (I.GT.0) THEN
        ISEED = I
      ELSE
        I = ISEED
      ENDIF
      RETURN
      END
C====================================================================
      SUBROUTINE POTENT1(MWALK,NWALKER,AMASS,X,V)
      IMPLICIT REAL*8 (A-H,O-Z)
      PARAMETER (OMEGA=2.0,OMEGA2=OMEGA**2)
      DIMENSION V(MWALK),X(MWALK)
C
C Compute the simple harmonic oscillator potential.
C
      DO 100 IW=1,NWALKER
        V(IW) = 0.5*AMASS*OMEGA2*X(IW)**2
  100 CONTINUE
      RETURN
      END
C====================================================================
      SUBROUTINE POTENT2(MWALK,NWALKER,AMASS,X,V)
      IMPLICIT REAL*8 (A-H,O-Z)
      PARAMETER (DD=0.17446,ALPHA=1.440,R0=1.4011)
      DIMENSION V(MWALK),X(MWALK)
C
C Compute the Morse oscillator potential with parameters
C selected to mimic vibrations in the hydrogen molecule; see S. Flugge,
C Practical Quantum Mechanics (Springer-Verlag, Berlin, 1974), pp. 68-71.
C Note that AMASS must be set to 918.05 (the reduced mass of two protons in
C atomic units) in the main program to reproduce H2.
C
      DO 100 IW=1,NWALKER
        R = (X(IW) - R0)/R0
        V(IW) = DD*(EXP(-2.0*ALPHA*R) - 2.0*EXP(-ALPHA*R))
  100 CONTINUE
      RETURN
      END
```

Appendix D

Bibliography

In this appendix we have listed in alphabetical order by topics all references cited in this book.

General References

1. M. Abramowitz and I. Stegun, eds., *Handbook of Mathematical Functions*, (Dover, New York, 1972).

2. A. D. Buckingham, "Permanent and Induced Molecular Moments and Long-Range Intermolecular Forces," *Advances in Chemical Physics* **12**, 107-142 (1967).

3. D. M. Ceperley and M. H. Kalos, "Quantum Many-Body Problems," in *Monte Carlo Methods in Statistical Physics*, 183-85 (Springer-Verlag, New York, 1979).

4. D. S. Chemla and J. Zyss, eds., *Nonlinear Optical Properties of Organic Molecules and Crystals* (Academic, New York, 1987).

5. E. Clementi, C. C. J. Roothaan and M. Yoshimine, "Accurate Analytical Self-Consistent Field Functions for Atoms. II. Lowest Configurations of the Neutral First Row Atoms," *Physical Review* **127**, 1618-20 (1962).

6. E. Clementi and C. Roetti, *Atomic Data and Nuclear Data Tables* **14**, 177-478 (1974).

7. W. Feller, *An Introduction to Probability Theory and its Applications* (Wiley, New York, 1968).

8. J. D. Graybeal, *Molecular Spectroscopy* (McGraw-Hill, New York, 1988).

9. J. M. Hammersley and D. C. Handscomb, *Monte Carlo Methods* (Chapman and Hall, London, 1964).

10. D. W. Heermann, *Computational Simulation Methods in Theoretical Physics* (Springer-Verlag, Berlin, 1986).

11. D. M. Hirst, *A Computational Approach to Chemistry* (Blackwell Scientific, London, 1990).

12. A. C. Hurley, *Introduction to the Electronic Theory of Small Molecules* (Academic, New York, 1976).

13. A. C. Hurley, *Electron Correlation in Small Molecules* (Academic, New York, 1976).

14. M. H. Kalos and P. A. Whitlock, *Monte Carlo Methods Volume 1: Basics* (Wiley, New York, 1986).

15. D. E. Knuth, *Seminumerical Algorithms*, 2nd ed., Vol. 2 of *The Art of Computer Programming* (Addison-Wesley, Reading, Mass., 1981)

16. G. C. Lie and E. Clementi, "Study of the Electronic Structure of Molecules. XXI. Correlation Energy Corrections as a Functional of the Hartree-Fock Den-

sity and its Application to the Hydrides of the Second Row Atoms," *Journal of Chemical Physics* **60**, 1275-87 (1974).

17. W. H. Press, B. P. Flannery, S. A. Teukolsky and W. T. Vetterling, *Numerical Recipes* (Cambridge University Press, 1986).

18. H. F. Schaefer, III, *The Electronic Structure of Atoms and Molecules. A Survey of Rigorous Quantum Mechanical Results*, (Addison-Wesley, Reading, 1972).

19. A. Szabo and N. S. Ostlund, *Modern Quantum Chemistry* (MacMillan Publishing Co., Inc., New York, 1982).

20. R. Y. Rubinstein, *Simulation and the Monte Carlo Method* (Wiley, New York, 1981).

21. A. Veillard and E. Clementi, "Correlation Energy in Atomic Systems. V. Degeneracy Effects for Second Row Atoms," *Journal of Chemical Physics* **49**, 2415-21 (1962).

22. M. Weissbluth, *Atoms and Molecules* (Academic Press, New York, 1978).

Review Articles

23. M. Caffarel, "Stochastic Methods in Quantum Mechanics," in *Numerical Determination of the Electronic Structure of Atoms, Diatomic and Polyatomic Molecules*, 85-105 (Kluwer Academic Publishers, 1989).

24. D. M. Ceperley and M. H. Kalos, "Quantum Many-Body Problems," in *Monte Carlo Methods in Statistical Physics*, 145-194 (Springer-Verlag, New York, 1979).

25. D. M. Ceperley, "A Review of Quantum Monte Carlo Methods and Results for Coulombic Systems," in *Monte Carlo Methods in Quantum Problems*, 47-57 (D. Reidel, Dordrecht, 1984).

26. B. L. Hammond, M. M. Soto, R. N. Barnett and W. A. Lester, Jr., "On Quantum Monte Carlo for the Electronic Structure of Molecules," *Journal of Molecular Structure (Theochem)* **234**, 525-38 (1991).

27. M. H. Kalos, "Optimization and the Many-Fermion Problem," in *Monte Carlo Methods in Quantum Problems*, 19-31 (D. Reidel, Dordrecht, 1984).

28. M. A. Lee and K. E. Schmidt, "Green's Function Monte Carlo," *Computers in Physics* **Mar/Apr**, 192-7 (1992).

29. W. A. Lester, Jr. and B. L. Hammond, "Quantum Monte Carlo for the Electronic Structure of Atoms and Molecules," *Annual Reviews of Physical Chemistry* **41**, 283-311 (1990).

30. J. W. Moskowitz and K. E. Schmidt, "Can Monte Carlo Methods Achieve Chemical Accuracy?" in *Monte Carlo Methods in Quantum Problems*, 59-70, (D. Reidel, Dordrecht, 1984).

31. P. J. Reynolds, J. Tobochnik and H. Gould, "Diffusion Quantum Monte Carlo," *Computers in Physics* **Nov/Dec**, 882-8 (1990).

32. K. E. Schmidt and M. H. Kalos, "Few- and Many-Fermion Problems," in *Monte Carlo Methods in Statistical Physics II*, 125-44 (Springer-Verlag, New York, 1984).

33. B. H. Wells, "Green's Function Monte Carlo Methods," in *Methods in Computational Chemistry* **1**, 311-50 (Plenum, New York, 1987).

Variational Monte Carlo

34. D. M. Ceperley, G. V. Chester and M. H. Kalos, "Monte Carlo Simulation of a Many-Fermion System," *Physical Review B* **16**, 3081-99 (1977).

35. H. L. Gordon, S. M. Rothstein and T. R. Proctor, "Efficient Variance-Reduction Transformations for the Simulation of a Ratio of Two Means: Application to Quantum Monte Carlo Simulations," *Journal of Computational Physics* **47**, 375-86 (1982).

36. J. Lee, "The Upper and Lower Bounds of the Ground State Energies Using the Variation Method," *American Journal of Physics* **55**, 1039-1040 (1987).

37. N. Metropolis, A. W. Rosenbluth, M. N. Rosenbluth, A. M. Teller and E. Teller, "Equations of State Calculations by Fast Computing Machines," *Journal of Chemical Physics* **21**, 1087-1092 (1953).

38. J. W. Moskowitz and M. H. Kalos, "A New Look at Correlations in Atomic and Molecular Systems. I. Applications of Fermion Monte Carlo Variational Method," *International Journal of Quantum Chemistry* **20**, 1107-119 (1981).

39. K. E. Schmidt and J. W. Moskowitz, "Correlated Monte Carlo Wave Functions for the Atoms He Through Ne," *Journal of Chemical Physics* **93**, 4172-4178 (1990).

40. C. J. Umrigar, K. G. Wilson and J. W. Wilkins, "A Method for Determining Many Body Wavefunctions," in *Computer Simulation Studies in Condensed Matter Physics: Recent Developments*, Springer Proceedings in Physics (Springer, Berlin, 1988).

41. T. Yoshida and K. Iguchi, "Variational Monte Carlo Method in the Connected Moments Expansion: H, H$^-$, Be, and Li$_2$," *Journal of Chemical Physics* **91**, 4249-53 (1989).

Diffusion Monte Carlo

42. J. B. Anderson, "Quantum Chemistry by Random Walk. H ^2P, H$_3^+$ D_{3h} $^1A_1$, H$_2$ $^3\Sigma_u^+$, H$_4$ $^1\Sigma_g^+$, Be ^1S," *Journal of Chemical Physics* **85**, 4121-27 (1976).

43. R. N. Barnett, P. J. Reynolds and W. A. Lester, Jr., "H + H$_2$ Reaction Barrier: A Fixed-Node Quantum Monte Carlo Study," *Journal of Chemical Physics* **82**, 2700-7 (1985).

44. R. N. Barnett, P. J. Reynolds and W. A. Lester, Jr., "Electron Affinity of Fluorine: A Quantum Monte Carlo Study," *Journal of Chemical Physics* **84**, 4992-6 (1986).

45. R. N. Barnett, P. J. Reynolds and W. A. Lester, Jr., "Is Quantum Monte Carlo Competitive: Lithium Hydride Test Case," *Journal of Physical Chemistry* **91**, 2004-5 (1987).

46. M. Caffarel and P. Claverie, "Development of a Pure Diffusion Quantum Monte Carlo Method Using a Full Generalized Feynman-Kac Formula. I. Formalism," *Journal of Chemical Physics* **88**, 1088-99 (1988).

47. M. Caffarel and P. Claverie, "Development of a Pure Diffusion Quantum Monte Carlo Method Using a Full Generalized Feynman-Kac Formula. II. Applications to Simple Systems," *Journal of Chemical Physics* **88**, 1100-9 (1988).

48. D. M. Ceperley, "The Statistical Error of Green's Function Monte Carlo," *Journal of Statistical Physics* **43**, 815-26 (1986).

49. J. D. Doll, "Monte Carlo Based Electronic Structure Techniques: Analysis and Applications," *Chemical Physics Letters* **81**, 335-38 (1981).

50. D. R. Garmer and J. B. Anderson, "Quantum Chemistry by Random Walk: Methane," *Journal of Chemical Physics* **86**, 4025-29 (1987).

51. D. R. Garmer and J. B. Anderson, "Quantum Chemistry by Random Walk: Application to the Potential Energy Surface for $F + H_2 \rightarrow HF + H$," *Journal of Chemical Physics* **86**, 7237-39 (1987).

52. R. Grimm and R. G. Storer, "A New Method for the Numerical Solution of the Schrödinger Equation," *Journal of Computational Physics* **4**, 230-49 (1969).

53. S.-Y. Huang, Z. Sun and W. A. Lester, Jr., "Optimized Trial Functions for Quantum Monte Carlo," *Journal of Chemical Physics* **92**, 597-602 (1990).

54. P. J. Reynolds, D. M. Ceperley, B. J. Alder and W. A. Lester, Jr., "Fixed-Node Quantum Monte Carlo for Molecules," *Journal of Chemical Physics* **77**, 5593-603 (1982).

55. P. J. Reynolds, J. Tobochnik and H. Gould, "Diffusion Quantum Monte Carlo," *Computers in Physics* **Nov/Dec**, 882-8 (1990).

56. S. M. Rothstein and J. Vrbik, "Statistical Error of Diffusion Monte Carlo," *Journal of Computational Physics* **74**, 127-42 (1988).

Exact Green's Function Monte Carlo

57. J. B. Anderson, "Simplified Sampling in Quantum Monte Carlo: Application to H_3^+," *Journal of Chemical Physics* **86**, 2839-43 (1987).

58. J. B. Anderson, C. A. Traynor and B. M. Boghosian, "Quantum Chemistry by Random Walk: Exact Treatment of Many-Electron Systems," *Journal Chemical Physics* **95**, 7418-25 (1991).

59. D. M. Arnow, M. H. Kalos, M. A. Lee and K. E. Schmidt, "Green's Function Monte Carlo for Few Fermion Problems," *Journal of Chemical Physics* **77**, 5562-72 (1982).

60. R. Bianchi, D. Bressanini, P. Cremaschi and G. Morosi, "Antisymmetry in the Quantum Monte Carlo Method with the A-Function Technique: H_2 b $^3\Sigma_u^+$, H_2 c $^3\Pi_u$, He 1 ^3S," *Journal of Chemical Physics* **98**, 7204-9 (1993).

61. D. M. Ceperley, "The Simulation of Quantum Systems with Random Walks: A New Algorithm for Charged Systems," *Journal of Computational Physics* **51**, 404-22 (1983).

62. D. M. Ceperley and B. J. Alder, "Quantum Monte Carlo for Molecules: Green's Functions and Nodal Release," *Journal of Chemical Physics* **81**, 5833-44 (1984).

63. D. L. Diedrich and J. B. Anderson, "An Accurate Quantum Monte Carlo Calculation of the Barrier Height for the Reaction $H + H_2 \rightarrow H_2 + H$," *Science* **258**, 786-88 (1992).

64. M. H. Kalos, "Monte Carlo Calculations of the Ground State of Three- and Four-Body Nuclei," *Physical Review* **128**, 1791-95 (1962).

65. M. H. Kalos, D. Levesque and L. Verlet, "Helium at Zero Temperature with Hard-Sphere and Other Forces," *Physical Review A* **9**, 2178-95 (1974).

66. M. A. Lee and K. E. Schmidt, "Green's Function Monte Carlo," *Computers in Physics* **Mar/Apr**, 192-7 (1992).

67. K. S. Liu, M. H. Kalos and G. V. Chester, "Quantum Hard Spheres in a Channel," *Physical Review A* **10**, 303-8 (1974).

68. D. W. Skinner, J. W. Moskowitz, M. A. Lee, P. A. Whitlock, and K. E. Schmidt, "The Solution of the Schrödinger Equation in Imaginary Time by Green's Function Monte Carlo. The Rigorous Sampling of the Attractive Coulomb Singularity," *Journal of Chemical Physics* **83**, 4668-72 (1985).

69. R. P. Subramaniam, M. A. Lee, K. E. Schmidt and J. W. Moskowitz, "Quantum Simulation of the Electronic Structure of Diatomic Molecules," *Journal of Chemical Physics* **97**, 2600-9 (1988).

70. S. Zhang and M. H. Kalos, "Exact Monte Carlo for Few-Electron Systems," *Physical Review Letters* **67**, 3074-77 (1991).

Trial Functions

71. J. D. Baker, J. D. Morgan, D. E. Freund and R. N. Hill, "Radius of Convergence and Analytic Behavior of the $1/Z$ Expansion," *Physical Review A* **43**, 1247-73 (1990).

72. H. Conroy, "Molecular Schrödinger Equation. I. One-Electron Solutions," *Journal of Chemical Physics* **41**, 1327-31 (1964).

73. H. Conroy, "Molecular Schrödinger Equation. IV. Results for One- and Two-Electron Systems," *Journal of Chemical Physics* **41**, 1341-51 (1964).

74. K. Frankowski and C. L. Pekeris, "Logarithmic Terms in the Wave Functions of the Ground State of Two Electron Atoms," *Physical Review* **146**, 46-49 (1966).

75. A. A. Frost, R. E. Kellogg and E. A. Curtis, "Local Energy Method in Electronic Energy Calculations," *Reviews of Modern Physics* **32**, 313-17 (1960).

76. Y. K. Ho, "Improved Hylleraas Calculations for Ground State Energies of Lithium Iso-Electronic Sequence," *International Journal of Quantum Chemistry* **20**, 1077-82 (1981).

77. E. A. Hylleraas, "Neue Berechnung der Energie des Heliums in Grand-austande, sowie des tiefsten terms von ortho-Helium," *Zeishrift für Physik* **54**, 347-366 (1929).

78. H. M. James and A. S. Coolidge, "Criteria of Goodness for Approximate Wave Functions," *Physical Review* **51**, 860-63 (1937).

79. R. J. Jastrow, "Many-Body Problem with Strong Forces," *Physical Review* **98**, 1479-84 (1955).

80. T. Kato, "On the Eigenfunctions of Many-Particle Systems in Quantum Mechanics," *Communications on Pure and Applied Mathematics* **10**, 151-177 (1957).

81. P. W. Kozlowski and L. Adamowicz, "An Effective Method for Generating Nonadiabatic Many-Body Wave Function Using Explicitly Correlated Gaussian-Type Functions," *Journal of Chemical Physics* **95**, 6681-98 (1991).

82. K. McDowell, "Assessing the Quality of a Wavefunction Using Quantum Monte Carlo," *International Journal of Quantum Chemistry: Quantum Chemistry Symposium* **15**. 177-81 (1981).

83. C. R. Meyers, C. J. Umrigar, J. P. Sethna and J. D. Morgan, "The Fock Expansion, Kato's Cusp Conditions, and the Exponential Ansatz," *Physical Review A* **44**, 5537-5546 (1991).

84. J. W. Moskowitz and K. E. Schmidt, "Correlated Monte Carlo Wave Functions for Some Cations and Anions of the First Row Atoms," *Journal of Chemical Physics* **97**, 3382-5 (1992).

85. C. L. Pekeris, "Ground State of Two-Electron Atoms," *Physical Review* **112**, 1649-58 (1958).

86. C. J. Umrigar, K. G. Wilson and J. W. Wilkins, "Optimized Trial Wave Functions for Quantum Monte Carlo Calculations," *Physical Review Letters* **60**, 1719-22 (1988).

87. C. J. Umrigar, M. P. Nightingale and K. J. Runge, "A Diffusion Monte Carlo Algorithm with Very Small Time-Step Errors," *Journal of Chemical Physics* **99**, 2865-90 (1993).

Excited States and Properties Other Than the Energy

88. S. A. Alexander, R. L. Coldwell, G. Aissing and A. J. Thakkar, "Calculation of Atomic and Molecular Properties Using Variational Monte Carlo Methods," *International Journal of Quantum Chemistry* **26**, 213-27 (1992).

89. R. N. Barnett, P. J. Reynolds and W. A. Lester, Jr., "Monte Carlo Determination of the Oscillator Strength and Excited State Lifetime for the Li $2\,^2S \rightarrow 2\,^2P$ Transition," *International Journal of Quantum Chemistry* **42**, 837-847 (1992).

90. R. N. Barnett, P. J. Reynolds and W. A. Lester, Jr., "Monte Carlo Algorithms for Expectation Values of Coordinate Operators," *Journal of Computational Physics* **96**, 258-76 (1991).

91. R. N. Barnett, P. J. Reynolds and W. A. Lester, Jr., "Computation of Transition Dipole Moments by Monte Carlo," *Journal of Chemical Physics* **96**, 2141-54 (1992).

92. P. Belohorec, S. M. Rothstein and J. Vrbik, "Infinitesimal Differential Diffusion Quantum Monte Carlo Study of CuH Spectroscopic Constants," *Journal of Chemical Physics* **98**, 6401-5 (1993).

93. B. Bernu, D. M. Ceperley and W. A. Lester, Jr., "The Calculation of Excited States with Quantum Monte Carlo. II. Vibrational Excited States," *Journal of Chemical Physics* **93**, 552-61 (1990).

94. J. S. Bowers, R. K. Prud'homme and R. S. Farinato, "An Assessment of the Padé-Laplace Method for Transient Electric Birefringence Decay Analysis," *Computers & Chemistry* **16**, 249-59 (1992).

95. M. Caffarel and D. M. Ceperley, "A Bayesian Analysis of Green's Function Monte Carlo Correlation Functions," *Journal of Chemical Physics* **97**, 8415-23 (1992).

96. M. Caffarel, M. Rérat and C. Pouchan, "Evaluating Dynamic Multipole Polarizabilities and van der Waals Dispersion Coefficients of Two-Electron Systems

with Quantum Monte Carlo. A Comparison with Some *Ab Initio* Calculations," *Physical Review A* **47**, 3704-17 (1993).

97. D. M. Ceperley and B. Bernu, "The Calculation of Excited-State Properties with Quantum Monte Carlo," *Journal of Chemical Physics* **89**, 6316-28 (1988).

98. D. F. Coker and R. O. Watts, "Quantum Simulation of Systems with Nodal Surfaces," *Molecular Physics* **58**, 1113-23 (1986).

99. A. L. L. East, S. M. Rothstein and J. Vrbik, "Sampling the Exact Electron Distribution by Diffusion Quantum Monte Carlo," *Journal of Chemical Physics* **89**, 4880-4 (1988).

100. R. M. Grimes, B. L. Hammond, P. J. Reynolds and W. A. Lester, Jr., "Quantum Monte Carlo Approach to Electronically Excited Molecules," *Journal of Chemical Physics* **85**, 4749-50 (1986).

101. W. Kolos and L. Wolniewicz, "Potential Energy Curves for the $X\ ^1\Sigma_g^+$, $b\ ^3\Sigma_u^+$ and $C\ ^1\Pi_u$ States of the Hydrogen Molecule," *Journal of Chemical Physics* **43**, 2429-41 (1965).

102. A. D. McLean and M. Yoshimine, "Molecular Properties Which Depend on the Square of Electronic Coordinates; H_2 and HNO," *Journal of Chemical Physics* **45**, 3676-81 (1966).

103. P. J. Reynolds, "Does Squaring the Quantum Monte Carlo Weights Give the Exact Quantum Probability Distribution?" *Journal of Chemical Physics* **92**, 2118-19 (1990).

104. P. J. Reynolds, R. N. Barnett, B. L. Hammond and W. A. Lester, Jr., "Molecular Physics and Chemistry Applications of Quantum Monte Carlo," *Journal of*

Statistical Physics **43**, 1017-26 (1986).

105. P. J. Reynolds, R. N. Barnett, B. L. Hammond, R. M. Grimes and W. A. Lester, Jr., "Quantum Chemistry by Quantum Monte Carlo: Beyond Ground-State Energy Calculations," *International Journal of Quantum Chemistry* **29**, 589-96 (1986).

106. P. J. Reynolds, M. Dupuis and W. A. Lester, Jr., "Quantum Monte Carlo Calculation of the Singlet-Triplet Splitting in Methylene," *Journal of Chemical Physics* **82**, 1983-90 (1985).

107. Z. Sun, W. A. Lester, Jr. and B. L. Hammond, "Correlated Sampling of Monte Carlo Derivatives with Iterative-Fixed Sampling," *Journal of Chemical Physics* **97**, 7585-89 (1992).

108. C. A. Traynor and J. B. Anderson, "Parallel Monte Carlo Calculations to Determine Energy Differences Among Similar Molecular Structures," *Chemical Physics Letters* **147**, 389-94 (1988).

109. C. J. Umrigar, "Two Aspects of Quantum Monte Carlo: Determination of Accurate Wavefunctions and Determination of Potential Energy Surfaces of Molecules," *International Journal of Quantum Chemistry: Quantum Chemistry Symposium* **23**, 217-30 (1989).

110. J. Vrbik, M. F. DePasquale and S. M. Rothstein, "Estimating the Relativistic Energy by Diffusion Quantum Monte Carlo," *Journal of Chemical Physics* **89**, 3784-87 (1988).

111. J. Vrbik, D. A. Lagare and S. M. Rothstein, "Infinitesimal Differential Diffusion Quantum Monte Carlo: Diatomic Molecular Properties," *Journal of Chemical*

Physics **92**, 1221-27 (1990).

112. J. Vrbik and S. M. Rothstein, "Infinitesimal Differential Diffusion Quantum Monte Carlo Study of Diatomic Vibrational Frequencies," *Journal of Chemical Physics* **96**, 2071-76 (1991).

113. B. H. Wells, "The Differential Green's Function Monte Carlo Method. The Dipole Moment of LiH," *Chemical Physics Letters* **115**, 89-94 (1985).

114. S. Zhang and M. H. Kalos, "Bilinear Quantum Monte Carlo: Expectations and Energy Differences," *Journal of Statistical Physics* **70**, 515-33 (1993).

Valence and Acceleration Methods

115. G. B. Bachelet, D. M. Ceperley, M. G. B. Chiocchetti, "Novel Pseudo-Hamiltonian for Quantum Monte Carlo Simulations," *Physical Review Letters* **62**, 2088-91 (1989).

116. G. G. Batrouni, G. R. Katz, A. S. Kronfeld, G. P. Lepage, B. Svetitsky and K. G. Wilson, "Langevin Simulations of Lattice Field Theories," *Physical Review D* **32**, 2736-47 (1985).

117. P. Belohorec, S. M. Rothstein and J. Vrbik, "Infinitesimal Differential Diffusion Quantum Monte Carlo Study of CuH Spectroscopic Constants," *Journal of Chemical Physics* **98**, 6401-5 (1993).

118. V. Bonifacic and S. Huzinaga, "Atomic and Molecular Calculations with the Model Potential Method," *Journal of Chemical Physics* **60**, 2779-86 (1974).

119. J. Carlson, J. W. Moskowitz and K. E. Schmidt, "Model Hamiltonians for Atomic and Molecular Systems," *Journal of Chemical Physics* **90**, 1003-6 (1989).

120. P. A. Christiansen, "Effective Potentials and Multiconfiguration Wave Functions in Quantum Monte Carlo Calculations," *Journal of Chemical Physics* **88**, 4867-70 (1988).

121. P. A. Christiansen and L. A. LaJohn, "Local Potential Error in Quantum Monte Carlo Calculations of the Mg Ionization Potential," *Chemical Physical Letters* **146**, 162-4 (1988).

122. P. A. Christiansen, "Core Polarization In Aluminum," *Journal of Physical Chemistry* **94**, 7865-67 (1990).

123. P. A. Christiansen, "Relativistic Effective Potentials in Transition Metal Quantum Monte Carlo Simulations," *Journal of Chemical Physics* **95**, 361-63 (1991).

124. H.-J. Flad, A. Savin and H. Preuss, "Reduction of the Computational Effort in Quantum Monte Carlo Calculations with Pseudopotentials Through a Change of the Projection Operators," *Journal of Chemical Physics* **97**, 459-63 (1992).

125. B. L. Hammond, P. J. Reynolds and W. A. Lester, Jr., "Valence Quantum Monte Carlo with *Ab Initio* Effective Core Potentials," *Journal of Chemical Physics* **87**, 1130-36 (1987).

126. B. L. Hammond, P. J. Reynolds and W. A. Lester, Jr. "Damped-Core Quantum Monte Carlo Method: Effective Treatment for Large-Z Systems," *Physical Review Letters* **61**, 2312-15 (1988).

127. M. M. Hurley and P. A. Christiansen, "Relativistic Effective Potentials in Quantum Monte Carlo Calculations," *Journal of Chemical Physics* **86**, 1069-70 (1987).

128. S., Huzinaga, L., Seijo, Z., Barandiaran, M. Klobukowski, "The *Ab Initio* Model Potential Method. Main Group Elements," *Journal of Chemical Physics* **86**, 2132-45 (1987).

129. L. R. Kahn, P. Baybutt and D. G. Truhlar, "*Ab Initio* Effective Core Potentials: Reduction of All-Electron Molecular Structure Calculations to Calculations Involving Only Valence Electrons," *Journal of Chemical Physics* **65**, 3826-53 (1976).

130. M. Krauss, W. J. Stevens, "Effective Potentials in Molecular Quantum Chemistry," *Annual Review Physical Chemistry* **35**, 357-85 (1984).

131. M. Lao and P. A. Christiansen, "Relativistic Effective Potential Quantum Monte Carlo Simulations for Ne," *Journal of Chemical Physics* **96**, 2161-63 (1992).

132. X.-P. Li, D. M. Ceperley and R. L. Martin, "Cohesive Energy of Silicon by the Green's-Function Monte Carlo Method," *Physical Review B* **44**, 10929-32 (1991).

133. L. Mitas, E. L. Shirley and D. M. Ceperley, "Nonlocal Pseudopotentials and Diffusion Monte Carlo," *Journal of Chemical Physics* **95**, 3467-75 (1991).

134. W. Müller, J. Flesch and W. Meyer, "Treatment of Intershell Correlation Effects in *Ab Initio* Calculations by Use of Core Polarization Potentials. Method and Application to Alkali and Alkaline Earth Atoms," *Journal of Chemical Physics* **80**, 3297-320 (1984).

135. J. C. Phillips and L. Kleinman, "New Method for Calculating Wave Functions in Crystals and Molecules," *Physical Review* **116**, 287-94 (1959).

136. P. J. Reynolds, "Overcoming the Large-Z Problem in Quantum Monte Carlo," *International Journal of Quantum Chemistry* **24**, 679-80 (1990).

137. W. J. Stevens, H. Basch and M. Krauss. "Compact Effective Core Potentials and Efficient Shared-Exponent Basis Sets for the First- and Second-Row Atoms," *Journal of Chemical Physics* **81**, 6026-33 (1984).

138. L. Szasz, *Pseudopotential Theory of Atoms and Molecules* (Wiley, New York, 1985).

139. C. J. Umrigar, "Accelerated Metropolis Method," *Physical Review Letters* **71**, 408-11 (1993).

140. T., Yoshida and K. Iguchi, "Quantum Monte Carlo Method with the Model Potential," *Journal of Chemical Physics* **88**, 1032-34 (1988).

141. T. Yoshida, Y. Mizushima and K. Iguchi, "Electron Affinity of Cl: A Model Potential Quantum Monte Carlo Study," *Journal of Chemical Physics* **89**, 5815-17 (1988).